Explainable AI with Python

Leonida Gianfagna • Antonio Di Cecco

Explainable AI with Python

Leonida Gianfagna
Cyber Guru
Rome, Italy

Antonio Di Cecco
School of AI Italia
Pescara, Italy

ISBN 978-3-030-68639-0   ISBN 978-3-030-68640-6   (eBook)
https://doi.org/10.1007/978-3-030-68640-6

This Springer imprint is published by the registered company Springer Nature Switzerland AG
The registered company address is: Gewerbestrasse 11, 6330 Cham, Switzerland

Contents

Chapter 1
The Landscape

*"Everyone knows that debugging is twice as hard as writing a
program in the first place.
So if you're as clever as you can be when you write it, how will
you ever debug it?"*
—Brian Kernighan

This chapter covers:

- What is Explainable AI in the context of Machine Learning?
- Why do we need Explainable AI?
- The big picture of how Explainable AI works

For our purposes we place the birth of AI with the seminal work of Alan Turing (1950) in which the author posed the question "Can machines think?" and the later famous mental experiment proposed by Searle called the *Chinese Room*.

The point is simple: suppose to have a "black-box"-based AI system that pretends to speak Chinese in the sense that it can receive questions in Chinese and provide answers in Chinese. Assume also that this agent may pass a Turing test that means it is indistinguishable from a real person that speaks Chinese. Would we be fine on saying that this AI system is capable of speaking Chinese as well? Or do we want more? Do we want the "black box" to explain itself clarifying some Chinese language grammar?

So, the root of Explainable AI was at the very beginning of Artificial Intelligence, albeit not in the current form as a specific discipline. The key to trust the system as a real Chinese speaker would be to make the system less "opaque" and "explainable" as a further requirement besides getting proper answers.

Jumping to our days, it is worth to mention the statement of GO champion Fan Hui commenting the famous 37th move of AlphaGo, the software developed by Google to play GO, that defeated in March 2016 the Korean champion Lee Sedol

© The Author(s), under exclusive license to Springer Nature Switzerland
AG 2021
L. Gianfagna, A. Di Cecco, *Explainable AI with Python*,
https://doi.org/10.1007/978-3-030-68640-6_1

with a historical result: "It's not a human move, I've never seen a man playing such a move" (Metz 2016). GO is known as a "computationally complex" game, more complex than chess, and before this result, the common understanding was that it was not a game suitable for a machine to play successfully. But for our purposes and to start this journey, we need to focus on Fan Hui's quoted statement. The GO champion could not make sense of the move even after having looked at all the match; he recognized it as brilliant, but he had no way to provide an explanation. So, we have an AI system (AlphaGo) that performed very well (defeating the GO champion), but no explanation of how it worked to win the game; that is where "Explainable AI" inside the wider Machine Learning and Artificial Intelligence starts to play a critical role.

Before presenting the full landscape, we will give some examples that are less sensationalistic but more practical in terms of understanding what we mean by the fact that most of the current Machine Learning models are "opaque" and not "explainable." And the "fil rouge" of the book will be to learn in practice leveraging different methods and how to make ML models explainable, that is, to answer the questions "What," "How," and "Why" on the results.

1.1 Examples of What Explainable AI Is

Explainable AI (aka XAI) is more than just a buzz word, but it is not easy to provide a definition that includes the different angles to look at the term. Basically speaking, XAI is a set of methods and tools that can be adopted to make ML models understandable to human beings in terms of providing explanations on the results provided by the ML models' elaboration.

We'll start with some examples to get into the context. In particular, we will go through three easy cases that will show different but fundamental aspects of Explainable AI to keep in mind for the rest of the book:

- The first one is about the *learning phase*.
- The second example is more on *knowledge discovery*.
- The third introduces the argument of *reliability and robustness* against external attacks to the ML model.

1.1.1 Learning Phase

One of the most brilliant successes of modern Deep Learning techniques against the traditional approach comes from computer vision. We can train a convolutional neural network (CNN) to understand the difference between different classes of labelled pictures. The applications are probably infinite: we can train a model to discriminate between different kinds of pneumonia RX pictures or teach it to translate sign language into speech. But are the results truly reliable?

Fig. 1.1 ML classification of wolves and dogs (Singh 2017)

Let's follow a famous toy task in which we have to classify pictures of wolves and dogs (Fig. 1.1).

After the training, the algorithm learned to distinguish the classes with remarkable accuracy: only a misclassification over 100 images! But if we use an Explainable AI method asking the model "Why have you predicted wolf?," the answer will be with a little of surprise "because there is snow!" (Ribeiro et al. 2016) (Fig. 1.2).

So maybe giving a model the ability to explain itself to humans can be an excellent idea. An expert in Machine/Deep Learning can immediately see the way a model goes wrong and make a better one. In this case, we can train the network with occlusions (part of the images covered) as possible augmentations (variations of the same pictures) for an easy solution.

1.1.2   Knowledge Discovery

The second case deals with Natural Language Processing (NLP) models like word embedding which can learn some sort of representation of the semantics of words. Words are embedded in a linear space as vectors, and making logic statement becomes simple as adding two vectors.

Man is to King as Woman is to Queen becomes a formula like:

$$Man - King = Woman - Queen$$

Fig. 1.2 Classification mistake. (**a**) Husky classified as wolf. (**b**) Explanation. (Ribeiro et al. 2016)

Fig. 1.3 ML classification of guitars and penguins with strange patterns (Nguyen et al. 2015)

Very nice indeed! But on the same dataset, we find misconceptions such as *Man is to Programmer as Woman is to Housekeeper*. As we say "garbage in garbage out," the data were biased, and the model has learned the bias. A good model must be Fair, and Fairness is also one of the goals of Explainable AI.

1.1.3   Reliability and Robustness

Now let's look at the picture below; you may see a guitar, a penguin, and two weird patterns (labelled again with Guitar and Penguin) with some numbers below. It is a result of an experiment in which state-of-the-art Deep Neural Network has been trained to recognize guitars and penguins (Fig. 1.3).

The numbers below each image are the confidence levels that Machine Learning system assigns to each recognition (e.g., say that the ML system is pretty sure that the first image is a guitar with 98.90% confidence level). But you may see that also the second image is recognized as a guitar with 99.99% CL and the fourth as a penguin with 99.99% CL. What is happening here?

This is an experiment conducted to fool the Deep Neural Network (DNN): the engineers maintained in the second and fourth images only the elements that the system used to recognize a guitar and a penguin and changed all the rest so that the system still "see" them like a guitar and a penguin. But these characteristics are not the ones that we, as humans, would use to do the same task; said in other way, these elements are not useful as an explanation to make sense of why and how the DNN recognizes some images (Nguyen et al. 2015).

1.1.4   What Have We Learnt from the Three Examples

As promised, let's critically think about the three examples above to see how they introduce different angles to look at Explainable AI capturing various aspects:

- *Example 1 about wolves classification*: accuracy is not enough as an estimator of a good Machine Learning model; without explanations we would not be able to discover the bad learning that caused the strong relation between snow and wolves.
- *Example 2 about Natural Language Processing*: there is the need for checking the associations against bias to make the process of knowledge discovery fair and more reliable. And as we will see in the following, knowledge discovery is among the main applications of XAI.
- *Example 3 about penguins*: this is trickier; in this case, the engineers did a kind of reverse engineering of the ML model to hack it, and we need to keep this in mind to see the relation between Explainable AI and making the ML models more robust against malicious attacks.

These experiments are trivial in terms of impacts on our life, but they can be easily generalized to cases in which a DNN is used to recognize tumor patterns or take financial decisions.

In these critical cases, we won't rely only on the outcome, but we will also need the rationale behind any decision or recommendation coming from the DNN, to check that the criteria are reliable and we can TRUST the system.

For an easy reminder, we'll require every Explainable AI model to be *F.A.S.T.* as in *Fair* and not negatively biased, *Accountable* on its decisions, *Secure* to outside malevolent hacking, and *Transparent* in its internals. Rethinking the examples, the second one needs more fairness and the last one more security.

This is precisely what Explainable AI (XAI) as emerging discipline inside Machine Learning tries to do: make the ML systems more transparent and interpretable to build trust and confidence in their adoption. To understand how XAI works, we need to do a step back and place it inside Machine Learning. Let's clarify from this very beginning that we are using terms interchangeably like interpretable and explainable. We will deep dive their meaning in the following starting from Sect. 1.4 of this chapter.

1.2   Machine Learning and XAI

Without going through a historical digression on how Machine Learning was born inside the broader context of AI, it is useful to recall some concepts for proper positioning of Explainable AI in this field and understand from a technical point of view how the need of explainability stands out.

Let's start with the figure below as a visual representation to place Machine Learning in the right landscape (Fig. 1.4).

Among the vast number of definitions for ML, we will base on this simple but effective one that captures very well the core:

> Machine Learning is the field of study that gives computers the ability to learn without being explicitly programmed. (A. Samuel 1959)

For our purposes, we need to focus on "without being explicitly programmed." In the old world of software, the solution to whatever problem was demanded to an algorithm. The existence of an algorithm guarantees by itself full explainability and full transparency of the system. Knowledge of the algorithm directly provides the explainability in terms of "Why" and "What" and "How" questions. Algorithms are a process or set of rules to be followed to produce an output given an input; they are not opaque to human understanding – all the knowledge around a problem is translated into a set of steps needed to produce the output.

The age of algorithms is being replaced by the *age of data* (Fig. 1.5); the current Machine Learning systems learn the rules from data during the learning phase and then produce the expected outputs based on the given inputs.

But you don't have directly the algorithmic steps that have been followed, and you might not be able to explain the reason for a specific output.

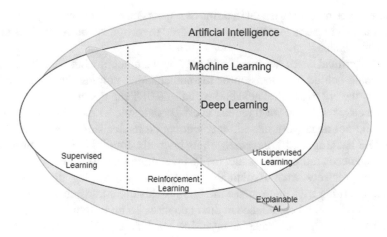

Fig. 1.4 Different areas of Artificial Intelligence in their mutual relations as Venn diagram

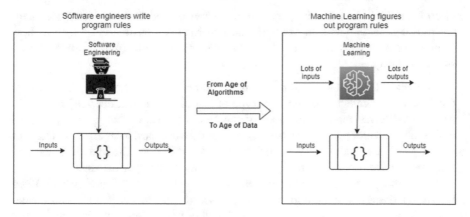

Fig. 1.5 Transition from standard software engineering driven by algorithms to software engineering driven by data

1.2.1   Machine Learning Tassonomy

Are all machine learning systems opaque by definition? No, it is not like that; let's have a quick categorization of Machine Learning systems before getting into details. There are three different main categories of Machine Learning systems based on the type of training that is needed (Fig. 1.4):

Supervised learning: the system learns a function to map inputs to outputs (A to B) based on a set of data that include the solutions (labels) on which the system is trained. At present supervised learning is the most used form of Machine Learning for the wide range of its possible applications. You can model as A to B correspondence spam filtering, ad-clicks prediction, stock prediction, sales prediction, language translation, and so on with a multitude of techniques, e.g., linear regressors, random trees, boosted trees, and neural networks.

Unsupervised learning: the training data are not labelled; they do not contain the solutions, and the system learns to find patterns autonomously (e.g., KMeans, PCA, TSNE, autoencoders). This is the part of ML that is more affine to GAI (General Artificial Intelligence) because we have a model that autonomously labels the data without any human intervention. A typical example of unsupervised learning is recommender systems like the ones used to suggest movies based on user's preferences.

Reinforcement learning: it is different from the two previous categories; there is no training on existing data. The approach here is to have an agent that performs actions in an environment and gets rewards that are specific to each action. The goal is to find a policy, i.e., a strategy, to maximize the rewards (e.g., Deep Learning networks and Monte Carlo tree search to play games like AlphaGo). We can think of an RL model like at the intersection of supervised and unsupervised systems for the model generates its own examples, so it learns in an unsupervised manner how to generate examples exploring the example's space and learning from them in a supervised way.

Looking at Fig. 1.4, *Deep Learning* is a subset of Machine Learning that does not fall in a unique category in terms of type of learning; the term deep refers specifically to the envisioned architecture of the neural networks that are implemented with multiple hidden layers making the neural network hard to interpret in terms of how it produces results. Deep Neural Networks (DNNs) are the Machine Learning systems that are producing the most successful results and performance.

Given the categories above (the three different types of learning and Deep Learning), there is not a unique mapping or set of rules to say that a specific category may need explainability more than the other in relation to the interpretability of the algorithms that belong to that category.

Explainable AI (XAI) is an emerging and transversal need (as pictured above) across the different AI domains. Let's make an example in supervised learning. Say we want to take a loan, and the bank says "NO." We want to know the reason, and the system cannot respond.

What if people eligible for a loan could be classified as in the following picture? (Fig. 1.6)

Here the axes are relevant features of the model like age and annual income of the borrower. As can you see, we can solve the problem with a linear classifier, and the model shows the suitable range of values for features to get the approval for the loan (Fig. 1.7).

But in a more complex model, we must face a trade-off problem. In Fig. 1.8 we may see the striking difference between the outcome of a simple linear classifier (Fig. 1.7) and a more sophisticated (most used) nonlinear one.

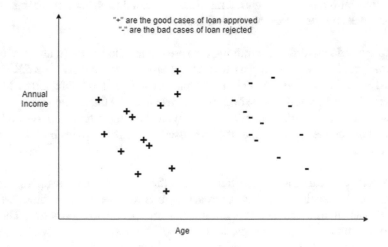

Fig. 1.6 Loan approval, good and bad cases

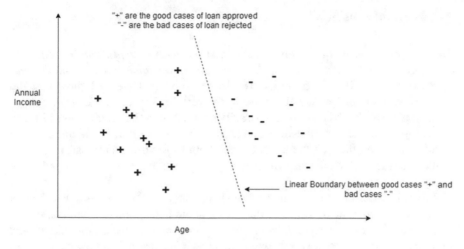

Fig. 1.7 Loan approval, good and bad cases with linear boundary

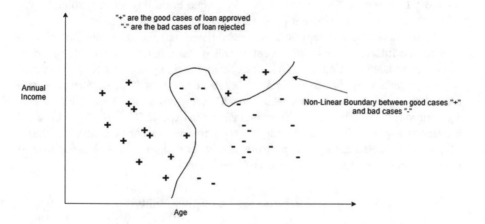

Fig. 1.8 Loan approval with nonlinear boundary

In this last case, it is not easy to explain the outcomes of the ML model in terms of approved and non-approved loans.

In linear models, you can easily say the effect of an increase or a decrease of a specific feature that is not generally possible for nonlinear cases. If you have a linear boundary to separate the two sets, you can explain how (in this specific case) age and annual income determine the approval or rejection of the loan. But with a nonlinear boundary, the two same features don't allow a straightforward interpretation of what's going on. We will go into details of how nonlinearity makes XAI harder starting from Chap. 2; for now, it is enough to note that the use of more complex models is unavoidable to achieve greater performance and inside Machine Learning, as we said, DNNs are the models that are most successful at all. This takes us to start demystifying two widespread beliefs.

1.2.2   Common Myths

The first common myth is that there is a strict need of explainability only for Deep Learning Systems (DNNs) that are opaque as constituted by many complex layers. This is only partially true in the sense that we will show that the need for explainability may also arise in the very basic Machine Learning models like regression and it is not necessarily coupled with the architecture of Deep Learning Systems. The question that may occur at this point is if there is any relation between the fact the DNN is seen both as the most successful ML systems in terms of performance (as noted previously) and the ones to need more explainability.

We will detail the answer in the following chapters, but it is useful to have a general idea right now of the standard answer in the field (somehow unjustified).

Indeed, *the second myth is that there is an unavoidable trade-off between Machine Learning system performance* (in terms of accuracy, i.e., how well images are recognized) and interpretability; you cannot get both at the same time – the better the ML system performs, the more it is opaque becoming less explainable. We can make a graph to visualize it (Fig. 1.9).

The two sets of points represent the qualitative trends for today and the expectation for the future. We will come back to this picture, but as for now it is already important to keep in mind that XAI is not for depressing performance to the advantage of explanations but to evolve the curve in terms of both performance and explainability at the same time, which means that given a model like DNN is performing very well today but not so explainable, the expectation is to move it toward increasing explainability but keeping (even improving) the performance. The trade-off between explainability and performance is on a single curve, but the overall trend is to push the curve for an overall improvement.

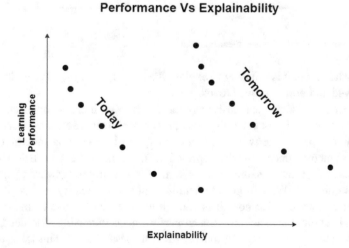

Fig. 1.9   Qualitative trend and relation of learning performance vs explainability

We will learn better across this book how these common beliefs have some validity, but they cannot be taken as general truths. Keeping this in mind, in the next two sections we will go into more details about the need of XAI beside the introductory examples and provide definitions of the terms we have been using to give them a reliable and operational meaning. As you may have noted, we have been using so far terms like explainable, interpretable, and "less opaque" interchangeably while they refer to slightly different aspects of the same concept to be clarified.

1.3   The Need for Explainable AI

The picture that is emerging from what is explained so far should allow to easily understand at this point why do we need Explainable AI. From one side it may be obvious to answer given the above examples, but it could be fair enough to use Machine Learning model to do a task given a high level of accuracy and performance without getting into details of making the model more transparent.

Let's try to understand the need of explainability in more general terms. As argued by Karim et al. (2018), a single metric like classification accuracy may not wholly describe and provide an answer to our real-world problem; getting the answer to "the what" might be useless without the addition of "the why," that is, the explanation of how the model gets the answer. In such cases the prediction itself is only a partial solution.

As a matter of fact, classification accuracy is not always good; in general, human field expertise in making appropriate metrics will be irreplaceable.

Science fiction is filled with Artificial General Intelligence agents that, no matter what is the task to perform with maximum accuracy, will take over to achieve it. "2001" had HAL, the spaceship's computer, deciding that it had to kill all the humans aboard because "this mission is too important for me to allow you to jeopardise it."

The three main applications that are often coupled with the prediction of a Machine Learning model and that introduce the need of explainability are (Du et al. 2019):

Model validation: it is connected to the properties of fairness, lack of bias, and privacy in relation to the ML model. Explanations are needed to check whether the Machine Learning model has been trained on a "Biased" dataset that may produce discriminations on a specific set of people. If a person is excluded from a loan, you need the possibility of looking into the black box to examine and produce the criteria that have been adopted for the decision. At the same time, the privacy of sensitive information is a must in specific cases (legal or medical among the others).

Model debugging: to guarantee reliability and robustness, the ML model should ensure some level of debugging that means the possibility to look behind the scenes into the machinery that produces the outputs. A small change in the inputs should not produce a huge change in the outputs to reduce the exposure to malicious attacks aimed at fooling the ML system and provide some level of robustness. Transparency and interpretability are needed to allow debugging in case of misbehavior and weird predictions.

Knowledge discovery: this is the most complex application to comment, being related to situations in which ML models are used not just to make predictions but to increase the understanding and knowledge of a specific process, event, or system. The extreme case that we will discuss further in the book is the adoption of ML models to gain scientific knowledge in which prediction is not enough without also providing explanations and causal relations. An infamous example that is helpful to understand the relation between explainability and trust is a rule-based model used to predict the mortality risks for patients with pneumonia. The unexpected result was that having asthma could decrease the risk of dying because of pneumonia. But the truth was that the patients with asthma were provided stronger medical treatments with better overall results; long story short, the need of explainability is deeply coupled with the level of trust we can have on a ML model.

It is easy to guess how the arguments connected to fairness, privacy, and trust represent fundamental factors that might strongly limit the adoption of ML systems in case explainability would not be guaranteed. We have not explicitly mentioned yet, in addition to the items above, the legal need of explainability because of regulations like GDPR in Europe that make explainability a must for AI, but this will be fully discussed in Chap. 8.

Are there any cases in which explainability might not be needed? The general answer is that the ML model may be treated just like black boxes only in the cases in which the model is not expected to produce any significant impact. And this is pretty evident if we look at the AI adoption speed in the consumer market in which there is a large diffusion of recommender systems and personal assistants to be compared to the still low adoption of AI in regulated industries.

1.4   Explainability and Interpretability: Different Words to Say the Same Thing or Not?

"If a lion could speak, we could not understand him" and not because of different languages but because of two different worlds or better two different "language games." We start with a quote from L. Wittgenstein to set the context and expectation of this section that is a bit more philosophical than the rest of the book. Wittgenstein worked in the domain of philosophy of language to investigate the conditions that make whatever statement understandable. For our purpose, we want to build ML models explainable, but we need to clarify before starting the journey into the real world of techniques what we mean or at least agree on what we mean for explainability and interpretability As in the case of lion, the language we use is strongly coupled with the world of our experience; the language is inherited from the world but also builds the world itself. So, we need to pay attention to the specific language we are searching for providing explanations about the opaque ML models. And we need to be sure about the domains, the "language games," in which the terms explainability and interpretability are used to avoid misunderstandings.

1.4.1   From World to Humans

Figure 1.10 makes more evident the concept: the layer "interpretability methods" lives between the black box and humans and two layers above the real world and the data with an increasing level of abstraction. We start from the bottom with grapes and wine that will be elements of the real case scenario we will handle in Chap. 3.

Interpretability methods should bridge the gap between predictions or decisions generated by the opaque ML models and humans to make them trust the predictions through explanation and interpretation.

It is not apparent to set the stage for the kind and level of acceptable explanations because of the source that is different from case to case (different ML models) and because of the target that is a generic human but with its own world, knowledge, and experience. There is not a unique or quantitative definition of interpretability, and in addition to this, usually, explainability and interpretability are used interchangeably.

Following Velez and Kim (2017), we define interpretability in the context of Machine Learning as "the ability to explain or to present in understandable terms to a human." The definition is fuzzy and non-operational, but it is used as a practical starting point for our analysis of the different meanings of interpretation and explanation. We want to state again that although these arguments may appear somehow abstract and philosophical, they are needed to set the stage before going into hands on through the different techniques.

1.4.2   Correlation Is Not Causation

To get the proper view on these ambiguities, we need to start from one of the applications of Explainable AI mentioned above, that is, knowledge discovery. In the age of data, the solid boundary between correlation and causation is fading away. There is no need to go into mathematical details to highlight the basic difference between the two terms; correlation is a statistical tool to check the connection between two or more items, to evaluate their coupling (the fact that change together).

But correlation doesn't imply causation that is something stronger and means that one variable causes the change in the other, one thing that makes the other to happen (cause and effect). If you see a correlation between doughnut sales and the number of homicides in a certain area, you are probably just seeing a coincidence

Fig. 1.10 From world to
humans through Machine
Learning

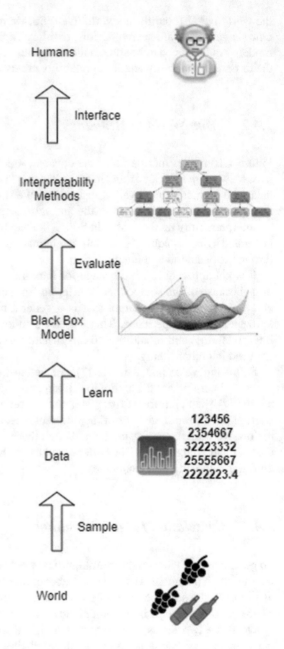

Humans

Interface

Interpretability
Methods

Evaluate

Black Box
Model

Learn

Data

Sample

World

without any underlying relation of causation between the two events. But when you
have a lot of data and the possibility to learn from them to make predictions, that is,
what Machine Learning does, you are almost relying on correlation to find patterns.
But you are not going into the direction of building new knowledge that requires
explanations, causal relations, and not just coincidences.

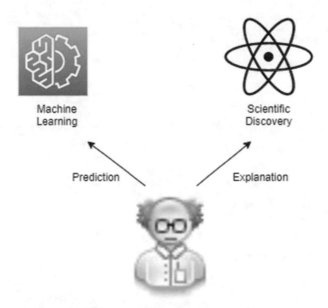

Fig. 1.11 Machine Learning, prediction vs explanation

This is an exciting stream of research inside the field of Machine Learning aimed at understanding if the classical scientific method based on models may be replaced by a brute analysis of the data to find patterns and generate predictions. For our purposes, it is important to emphasize again the concept that making predictions with Machine Learning is basically doing correlation, and these predictions need to be *interpreted and explained* to generate knowledge (Fig. 1.11).

We explicitly said "interpreted and explained" because these are two different actions. Most of the techniques and tools that we will study and discuss through this book are tools aimed at providing interpretations of the opaque ML models, but this could not be enough to get explanations. Interpretations are an element of an explanation but do not exhaust it.

To get a an analogy from science, the same theory, for example, quantum mechanics, can be used to build predictions that work well without understanding absolutely anything of it; then, you may come to interpretations that are not just one but many (Copenhagen interpretations with collapsing wave functions, many-worlds interpretation) which are tested by predictions in a continuous loop to generate a full theory – *an explanation that is composed of interpretations expressed through formalisms and predictions* (Fig. 1.12).

The same argument can be rephrased with a more practical and tangible example on ML. The picture below shows how to generate different good models to learn the risk associated with granting a loan to a customer depending on income and interest rate variables (Fig. 1.13).

If we model the error as a mountain landscape, each choice of parameters makes a different model, and we prefer those that minimize the error (loss) function. So,

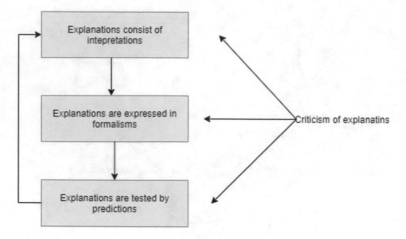

Fig. 1.12 Explanations decomposed (Deutsch 1998)

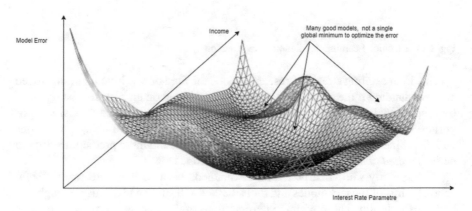

Fig. 1.13 An illustration of the error surface of Machine Learning model

every local minimum on the landscape of the error function makes, to some performance, a good choice of the model, and each different model would generate a different set of interpretations. But you don't have a global model that comprises the overall phenomenon, the kind of explanation we expect from a scientific theory.

1.4.3   So What Is the Difference Between Interpretability and Explainability?

To provide a further visual example of this distinction between interpretability and explainability, let's think about the boiling water; the temperature increases with time steadily until the boiling point after which it will stay stable. If you just rely on

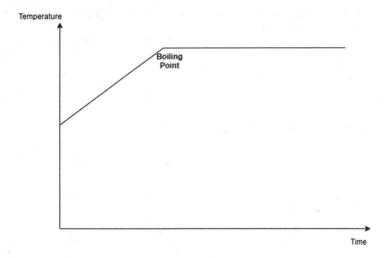

Fig. 1.14 Water phase transition diagram with two different trends for temperature before and after the boiling point

data before the boiling point, the obvious prediction with the related interpretation would be that temperature rises continuously. Another interpretation may make sense of data taken after the boiling point with a steady temperature.

But if you search for a full explanation, a full theory of water "changing state," this is something deeper that exceeds the single good interpretations and predictions, in the two different regimes. ML would be good at predicting the linear trend and the flat temperature after the boiling point, but the physics of the phase transition would not be explainable (Fig. 1.14).

Interpretability would be to understand how the ML systems predict temperature with passing time in the normal regime; explainability would be to have a ML model that takes into account also the changing state that is a global understanding of the phenomenon more related to the application of knowledge discovery already mentioned.

To summarize, with the risk of an oversimplification of the discussion above but getting the core, *we will consider interpretability as the possibility of understanding the mechanics of a Machine Learning model but not necessarily knowing why.*

To provide an operational approach on how to distinguish interpretability from explainability, we summarized what is stated above with a table that differentiates the two terms relying on the different questions that may be answered in each of the two scopes (Table 1.1).

Explainability is for us something more, in terms of being able to answer questions about what happens in case of new data, "What if I do x, does it affect the probability of y" and counterfactual cases to know what would have changed is some features (or values) would not have occurred. Explainability is a theory that deals also with unobserved facts toward a global theory, while interpretability is limited to making sense of what is already present and evident.

Table 1.1 Difference between interpretability and explainability in terms of the questions to answer for the two different scopes

Question	Interpretability	Explainability
Which are the most important features that are adopted to generate the prediction or classification?	✓	✓
How much the output depends on small changes in the input?	✓	✓
Is the model relying on a good range of data to select the most important features?	✓	✓
What are the criteria adopted to come across the decision?	✓	✓
How would the output change if we put different values in a feature not present in the data?	✗	✓
What would happen to the output if some feature or data had not occurred?	✗	✓

As stated by Gilpin et al. (2018): "We take the stance that interpretability alone is insufficient. For humans to trust black-box methods, we need explainability – models that can summarise the reasons for neural network behaviour, gain the trust of users, or produce insights about the causes of their decisions. *Explainable models are interpretable by default, but the reverse is not always true.*"

Keeping in mind these distinctions, we will have our path through the different techniques to make Machine Learning models more explainable. In most of the cases, the methods and systems we will present are aimed at providing interpretability and not explainability as we have been discussing in this section, and we will make this evident.

And it won't be important for most of the time to come back again on the difference between interpretability and explainability until the end of the book in which we will touch the argument of knowledge discovery (Chap. 6 focused on how we may do science with ML and XAI) again and try to present a framework for a standard approach to AI (Chap. 8) in which this distinction will rise again.

1.5   Making Machine Learning Systems Explainable

The landscape of Explainable AI should be clear at this point; we provided a high-level description of what Explainable AI is and why it is needed in the broader context of Machine Learning. Also, we tried to present better and clarify the terms and buzzwords that are used in this field. This section aims to depict a global map of the Explainable AI (XAI) system and process that may be used to get in on shot the big picture and as orientation for the rest of this work.

1.5.1   The XAI Flow

The best starting point is to locate the XAI inside the classical Machine Learning pipeline that briefly consists of three phases (we are not getting into details of data preparation and optimization for our purposes, Fig. 1.15):

1. Training data
2. Machine Learning process
3. Learned function that generates prediction, decision, or recommendation

The main point of XAI is to make sense of the output producing explanations and interpretations that can be understood by humans, that is, to make the model explainable and provide an explanation interface open to the users.

The two blocks inside the red ellipse – the explainable model and explainable interface – represent the core content of this book. They are further expanded in Fig. 1.16. Note that if exploded, the two blocks become methods and systems, and not a system only. This means that we are not making a ML system explainable, just changing his inner components; but most of the time, the ML is left untouched, and we make sense of it from the external through the proper techniques.

As per the picture below, the XAI mental model is a flow that, given a ML model, provides the proper options and techniques to make it explainable.

The map now should be easily readable, but let's just emphasize the main points. We have a given ML process (existing or built from scratch) that is providing outputs through the learned function, and we need to make it explainable. The "explainable model" of Fig. 1.15 is exploded in Fig. 1.16 into the different techniques and approaches that, given the original Learned Function, drive to the explainable human interface and the human-readable XAI metrics. Note that there are two main decision points in the flow: in the first one, the "human" may choose the main XAI approach to adopt, that is, an agnostic approach, or a model-dependent one. These techniques will be deeply dived starting from Chap. 2; for now it is enough to get the main points below:

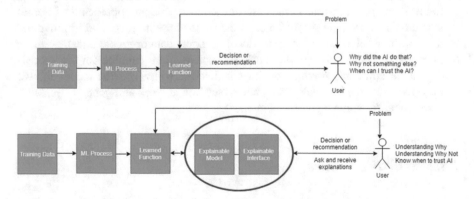

Fig. 1.15 Machine Learning pipeline with focus on XAI, blocks inside the ellipse

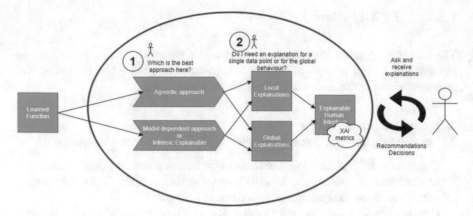

Fig. 1.16 Different approaches to make a ML model explainable

- *Agnostic approach* means that the XAI works with the ML model as a "black box" without assuming any knowledge of the internals to produce explanations.
- *Model-dependent approach or intrinsic explainable* means that knowledge of the ML model internals are used to produce explanations. As sub-case we put here also the "intrinsic explainable" models (we will look at them in Chap. 3) in which the model parameters provide explanations directly.

The decision point number "2" is about the choice of a global vs local explanation: this depends from case to case (but again we will detail that starting from Chap. 2), and it is related to the need of getting a global explanation for the full ML model behavior or only for a specific subset of data.

Then we can do cycles until the produced explanations (XAI metrics) are satisfactory (Fig. 1.17).

Figure 1.17 shows the two main types of loops to provide feedback to the system and improve explanations:

Black-box perturbations is to feed the model with artificial data, even strange data, to test its response, while in the case of model-dependent approach the internal structure of the model might be changed to see how it reacts and improve the level of the produced explanations (local for some data or global ones).

Change internals and observe system's reaction is more suitable for the model-dependent approach in which we have full access to the specific model and we may play with the internal parameters to see how it work and generate explanations (local for some data or global ones).

1.5.2 The Big Picture

The spirit of the book is not only to provide techniques to deal with XAI if needed and produce model interpretations but mostly to complement the practical examples with critical thinking to understand the real reasons behind the adopted techniques

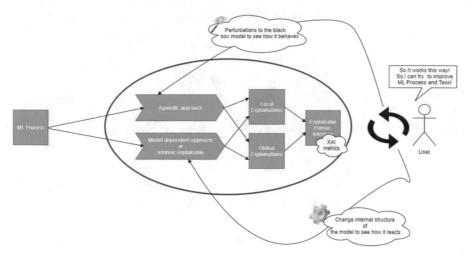

Fig. 1.17 Different approaches to make a ML model explainable with focus on the feedback from the external to further improve XAI

and the urge to adopt them and to avoid false expectations. The technology rapidly changes; so in terms of trading off we prefer to try to share a mindset, a way of thinking with practical methods instead of just a list of references to XAI methods and tools. Each XAI technique has its own domain of applicability and limitations; every technique (global or local, agnostic or model-dependent as in Fig. 1.17) needs to be carefully chosen and tailored toward our objective. We will use our mental model across the book to keep always the focus on the global XAI picture and quickly position the specific technique we will deep dive into the proper context. So, the figure below (same flow but without annotations to keep it simple) will appear again and again to see where we are during our XAI journey.

While Fig. 1.18 represents the high-level internals to make a ML model explainable, it might be useful to expand the flow to have a map of the book with pointers to the different chapters (Fig. 1.19).

As discussed, there are different levels of possible explanations that could be required, and critical thinking about these levels will provide the reader a way to move toward a standardized path of certification of a XAI system if needed (at the bottom of the flow).

To have another look at the flow, we will have another chapter (Chap. 2) still focused on theory and landscape before putting hands-on code to use XAI methods on real-life scenarios. Real-life scenarios here mean to put in the shoes of a XAI scientist who is asked to provide explanations on the predictions provided by a ML model. We will learn how to use these methods to answer "What," "Why," and "How" questions on the predictions from the ML models.

We will also deep dive into the robustness and security of ML models from a XAI perspective to prevent (or to be aware) attacks aimed at fooling the ML models to change their predictions. The ultimate goal is to look at XAI from different perspectives to successfully deal with problems coming from real life to trust and certify ML models as explainable.

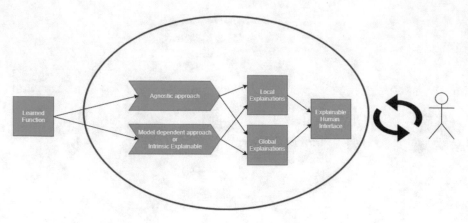

Fig. 1.18 XAI main flow

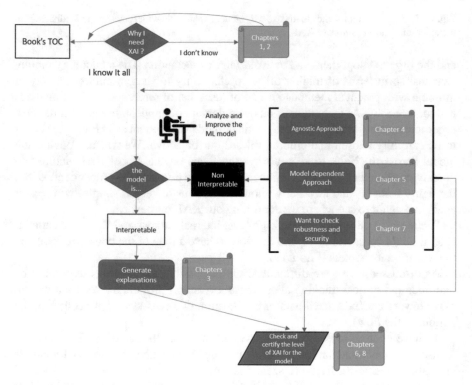

Fig. 1.19 Visual map of the main concepts and topics across the book

1.6 Do We Really Need to Make Machine Learning Models Explainable?

In this section we want to play the role of a devil's advocate and challenge the real need of explainability beside the arguments already discussed before jumping into the real world of XAI methods of the next chapters.

The mantra around XAI is that to trust Machine Learning we must be able to explain how the decisions or predictions have been generated. But let's think again about the core of Machine Learning; basically we are talking about systems that solve problems relying on learning by examples (experience) instead of algorithms (explanation of how things work).

And the best cases in which Machine Learning shows its full strength are the cases in which providing all the rules is not feasible (e.g., describe how to recognize a cat with an algorithm instead of training the system with millions of cat images).

Thus, could we say that you might get explainability at the price of losing the power of Machine Learning? This would mean that you will get explainable models only in the cases in which Machine Learning is not producing any huge benefit having already a complete algorithm to solve the problem.

To make these arguments more practical, let's suppose that we are determined to get the secrets of how our favorite football player kicks the ball (Fig. 1.20).

Would you try to get the physics of the kick or try to learn by experience looking at him and doing the same again and again? I would guess that the physics model would be useless and you would learn by experience. In this analogy relying on the physics model to learn how to kick is the same as using XAI to understand how a Deep Machine Learning model works.

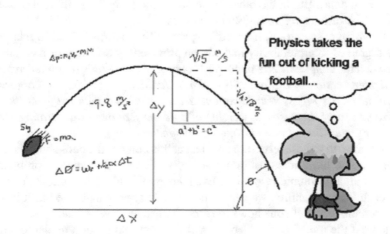

Fig. 1.20 Is knowing physics really necessary to play football? (ChaosKomori 2003)

If you remember one of the original streams of AI, the idea was to make neural networks mimic how the human brain learns and works. And up to some extent, Deep Neural Networks are doing exactly that: learning from experience to do complex tasks that does not necessarily mean having an "acceptable" explanation of how to perform these tasks.

And this is totally acceptable for humans in the sense that you would never rely on a deep dive in neuroscience to get the neurons' activity and provide such an explanation of how a person is performing a mathematical task. In case of a student, in order to check if he is learning math, you would test him with questions or problems that the student should be able to solve given the required level of knowledge. But you would not care about "how" the student derived his answers. And this is exactly what happens when you transfer a Deep NN to work on test data after training.

So is this to say that XAI is not really needed and close the book? Not at all; the message is again to rely on critical thinking to understand the context. Much of misunderstanding comes from not knowing or from misinterpreting the background of your problem and the related needs.

Let's get back to the main applications that need XAI: model validation, model debugging, and knowledge discovery. For the last one, knowledge discovery, there is no sense in having predictions with Machine Learning without explanations. In this case we are trying to understand a phenomenon by relying on a brute Machine Learning approach on data and without any abstract modeling. Also, assuming we have the right predictions, it won't be any progress in the knowledge without explanations (because of the difference between correlation and causation).

In this specific case, we strongly need details on how the system works also at the risk of decreasing the interpretability, because we want to answer questions about system behavior that are in the scope of new data and counterfactual examples (unobserved data). But in the other two cases, we are targeting different properties: fairness, lack of bias, reliability, and robustness.

What we may search for, if not full explainability in terms of causal relations, is some degree of interpretability that can be provided by a variety of methods (discussed in later chapters). *The important concept to get here is that interpretability, as a lighter version of explainability, can be fulfilled also without an algorithmic equivalent of the ML system (complete set of rules) but with artifacts like local approximation of the systems, weights of different features in generating the outcomes, or generation of rules through decomposition.*

It is normal that these terms are not clear at this point of the book, but it is enough to get now the basic idea: using a black-box ML model to solve a problem that doesn't fit an algorithmic resolution. Even though this black box might not be suitable for full explainability, we can do things (like approximate the black box with an interpretable model somehow) to reach the needed level of interpretability and start to trust the model itself. Also we will see how asking for a model to be interpretable and/or explainable will help on building a better model as well with best practices that on one side may guarantee the explanations and on the other avoid problems with overfitting and bias among the others.

1.7 Summary

- Understand what is meant by "Machine Learning systems are getting 'opaque' to human understanding."
- Use practical examples to show how the trust in ML may be reduced without the possibility of answering "Why" questions about the output and adopted criteria.
- Explainable AI is an emerging discipline inside Machine Learning to make ML models more interpretable. You need XAI to generate trust on ML.
- Explainable AI might be critical for the application of ML in regulated industries like finance and health. Without XAI ML application might be strongly reduced in scope.
- Distinguish and understand the primary applications of XAI in terms of model validation, model debugging, and knowledge discovery.
- Explainability and interpretability are often used interchangeably, but they mean different things.

In the next chapter, we will start to detail the different approaches to XAI depending on the specific context (which ML model needs to be interpreted) and needs (which level of explanation I'm looking for).

References

ChaosKomori. (2003). *The physics of football*. DeviantArt. Available at https://www.deviantart.com/chaoskomori/art/The-Physics-of-Football-1870988.

Deutsch, D. (1998). *The fabric of reality*. London: Penguin.

Du, M., Liu, N., & Hu, X. (2019). Techniques for interpretable machine learning. *Communications of the ACM, 63*(1), 68–77.

Gilpin, L. H., Bau, D., Yuan, B. Z., Bajwa, A., Specter, M., & Kagal, L. (2018). Explaining explanations: An overview of interpretability of machine learning. In *2018 IEEE 5th international conference on data science and advanced analytics (DSAA)* (pp. 80–89). IEEE.

Karim, A., Mishra, A., Newton, M. A., & Sattar, A. (2018). Machine learning interpretability: A science rather than a tool. *arXiv preprint arXiv:1807.06722*.

Metz, C. (2016). *How Google's AI viewed the move no human could*. Available at Understand https://www.wired.com/2016/03/googles-ai-viewed-the-move-no-human-understand/.

Nguyen, A., Yosinski, J., & Clune, J. (2015). Deep neural networks are easily fooled: High confidence predictions for unrecognizable images. In *Proceedings of the IEEE conference on computer vision and pattern recognition* (pp. 427–436).

Ribeiro, M. T., Singh, S., & Guestrin, C. (2016). "Why should I trust you?" Explaining the predictions of any classifier. In *Proceedings of the 22nd ACM SIGKDD international conference on knowledge discovery and data mining* (pp. 1135–1144).

Samuel, A. L. (1959). Some studies in machine learning using the game of checkers. *IBM Journal of Research and Development, 3*(3), 210–229.

Singh, S. (2017). *Explaining black-box machine learning predictions*. Presented at #H2OWorld 2017 in Mountain View, CA. Available at https://youtu.be/TBJqgvXYhfo.

Turing, A. M. (1950). Computing machinery and intelligence-AM Turing. *Mind, 59*(236), 433–460.

Doshi-Velez, F. & Kim, B. (2017). Towards a rigorous science of interpretable machine learning. arXiv preprint arXiv:1702.08608.

Chapter 2
Explainable AI: Needs, Opportunities, and Challenges

"One could put the whole sense of the book perhaps in these words: What can be said at all, can be said clearly; and whereof one cannot speak, thereof one must be silent."
—Ludwig Wittgenstein

This chapter covers:

- What is an explanation, and how to evaluate it?
- Subtleties on the need of making a ML model explainable.
- High-level overview of the different XAI methods and properties.

This chapter is a bridge between the high-level overview of XAI presented in Chap. 1 and the hands-on work with XAI methods that we will start in Chap. 3. The chapter will introduce a series of key concepts and a more complete terminology as you will find in literature and papers.

The examples in Chap. 1 made evident the cases in which XAI needs to be coupled with ML model predictions to make them useful and trusted. We also saw what is really meant by words like explainability and interpretability.

The core of this chapter is to provide a general presentation of XAI methods with the proper taxonomy. In order to do this, we will rely on a practical example in which the goal is to forecast sales of a product depending on the age of the customers.

Before jumping to a general presentation of XAI methods, we will start this chapter with some discussion about how to evaluate an explanation from a human standpoint and the role that humans may play into ML and XAI pipeline we presented in the previous chapter.

© The Author(s), under exclusive license to Springer Nature Switzerland AG 2021
L. Gianfagna, A. Di Cecco, *Explainable AI with Python*,
https://doi.org/10.1007/978-3-030-68640-6_2

2.1   Human in the Loop

In Chap. 1, we briefly touched on the ambiguities that emerge when using different words like explainability and interpretability. The position adopted in this book is that explainability is a stronger request than interpretability. Interpretability is a first step to gain explainability in which methods are adopted to get some hints on how the ML model is producing a certain output. But explainability requires a full understanding of the ML model, the possibility to answer "Why" questions at the point of being able to anticipate the outcomes.

The tricky point in this path is how to check that an explanation is good enough and who is the audience for the "good enough" explanations. You may easily understand that a good explanation for a ML practitioner in terms of technical details might not work at all for people coming from outside of ML world. People not in the field might not increase their trust in the model with such kind of technical assessment only.

2.1.1   Centaur XAI Systems

If you think about a XAI system as a "Centaur" combining a Machine Learning model with a human trying making sense of the explanations coming from the model, we can use it as a model for the "human in the loop"-XAI paradigm.

In Fig. 2.1 we have portrayed a typical task such as an AI classifier which outputs a "dog vs wolf" classification with some probability.

If the classifier is not confident enough, it can ask for help from the human. The human can add new annotations to the pictures or make fine-tunings to the model. This form of collaboration between AI systems and humans is aimed at further

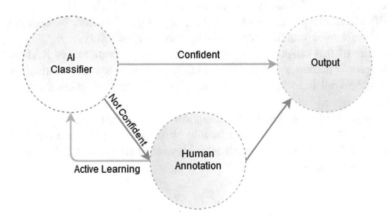

Fig. 2.1 The human in the loop improves the performance of an AI classifier taking part in the training process

improvement of performance. Of course, humans can use active learning even in the earlier steps by choosing a best-suited model for the classifier or building smarter features the model can use in the training process.

The typical example to understand this is to think about what has been happening in chess. The common belief that followed the victory of IBM's Deep Blue against Gary Kasparov in 1997 was that there was not further room for humans to play chess against machines. But it was the same Kasparov to raise the doubt, 1 year later in 1998, that maybe a human might not defeat an AI system anymore but a "human in the loop" could successfully collaborate with a machine to play chess against a human or a machine with better performance.

Long story short, freestyle chess tournaments were organized, and the results of these Centaur systems (humans plus AI) have been impressive against the common belief: a strong computer plus a strong human consistently beat the strongest computers on their own. But it is even more than that; it has been shown how three weak computers plus two young chess amateur players won against Hydra, one of the most powerful chess AI at the time. Where the three weak computers had different recommendations, the humans may interact with the system to do further analysis and take a decision. In a way, the interaction between humans and machine models is more than the mere sum of the parts. The use of Machine Learning models exposes some human planning ability that artificial machine models still lack.

This example that comes from chess can be extended in a more general way. *Putting humans in the loop is not only needed for XAI but also to achieve better results.* This can be effectively summarized *in a variation of Pareto's 80:20 rule.*

The usual call for Pareto principle (V. Pareto was an Italian economist that set the 80/20 rule in 1896 at the University of Lausanne) is that for a variety of events, the 80% of effects usually come from 20% of the causes, and it is usually adopted as a mantra in the corporations to say that 80% of sales come from 20% of clients or that 80% of the software quality is achieved by fixing 20% of the most common defects. The mathematical root of this outcome is to be searched in the underlying power-law probability distribution that produces the phenomena.

In this case, an ideal Machine Learning system is one in which 80% of the results are AI-driven, but in order to further boost accuracy, 20% of the efforts should come from humans (Fig. 2.2).

Humans have an active role not only as in the XAI domain to receive and produce explanations but also as active players to enhance the overall performance as in Centaur collaboration models. *Machines are good at providing answers, but humans are often better on finding the right questions* to take critical decisions or to interpret results for rare cases outside of the learning dataset.

Also, models are shockingly weak when it comes to answer questions such as "What is more beautiful?" or "What is the right thing to do?" or to answer those

Fig. 2.2 The Pareto
principle suggests that an
efficient ML system must
have a 20% of creative
(human) effort

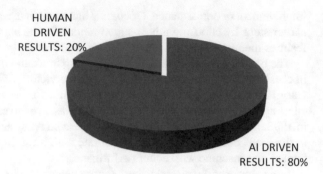

cases arising from explainability. For example, it is an open question how to create a model that will be fair and not affected by bias like the "Man is to Woman as Programmer is to … Home maker" problem we talked about in Chap. 1.

But pragmatically a model can ask human collaborators "What answers are Fair and which ones are not Fair?" and then attempt to learn from the answers. For such reasons in XAI, humans can contribute to all four F.A.S.T. aspects of explaining a model (remember that F.A.S.T. acronym is a reminder for the attributes' Fairness, Accountability, Security, and Transparency).

Do you remember the common belief about the need to trade-off between explainability and performance we already talked about? This is another angle to demystify it. Performance comes together with "humans in the loop" boosting performance and explainability at the same time.

For example using some field expertise, we can build smarter features making simpler and more explainable models while boosting performance.

Back to the specific XAI scope but having in mind this pattern of collaboration, XAI systems put a "human in the loop" to make sense of the output by adding explanations and interpretations that can be understood by the humans with an interface open to the users.

2.1.2 XAI Evaluation from "Human in the Loop Perspective"

As outlined by Gilpin et al. (2018), an explanation can be assessed by two main features: its interpretability and its completeness.

The main objective of *interpretability* is to provide a set of descriptions that allows a person to understand what the ML model is doing up to the extent of gaining meaningful knowledge and trust in the system, according to a person's specific needs.

Completeness is the accuracy of the description of the systems up to the possibility of anticipating the results of the model. So a description of the model is complete if it can distill all the knowledge of the model in human-understandable language.

In this sense, it could be difficult to trade-off between interpretability and completeness because the most interpretable explanations are usually simple and a description of the model that is complete can be as complex as the model itself. As argued by Gilpin: "Explanation methods should not be evaluated on a single point on this trade-off, but according to how they behave on the curve from maximum interpretability to maximum completeness" (Gilpin et al. 2018).

This curve depends explicitly on the specific human that has been put in the loop to find the most suitable trade-off for the specific scenario under examination. Among the different options, we follow the approach provided by Doshi-Velez and Kim (2017) to summarize the different categories of explainability based on the role played by the human in the evaluation (Fig. 2.3):

- *Application-grounded evaluation* (real humans – real tasks): this approach regards humans trying to complete real tasks relying on the produced explanations coming from the application of XAI to the ML system. In this case the assumption is that the human is an expert in the domain of the task (e.g., a doctor doing a diagnosis for a specific disease). The evaluation is conducted based on the results achieved by the human (better quality, less errors) relying on the explanations.
- *Human-grounded evaluation* (real humans – simple tasks): in this case we don't have domain experts but just lay humans to judge the explanations. The evaluation approach depends on the quality of explanation independently from the associated prediction. For example, the human may be provided an explanation and an input and is asked to simulate the model's output (regardless of the true output). We can think of a credit approval system, for example, where a loan officer must explain to a bank customer why their loan has been refused. In this case the system can show to the costumer some minimal set of pertinent negative features (e.g., number of accounts or age) that once altered would change their eligibility into a different state.

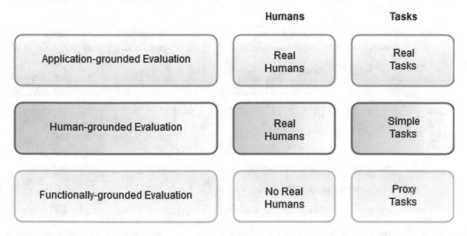

Fig. 2.3 Categories of explainability based on the role played by the human in the evaluation

- *Functionally grounded evaluation* (no real humans – proxy tasks): this kind of evaluation is usually adopted when there are already a set of models that have been validated by human-grounded experiments as explainable which are then used as a proxy for the real model. They are used when there is no access to humans to perform evaluation. So back to the example of the credit approval system, in this case we don't have humans to judge the explanations, but we use an interpretable model that should act as a proxy to evaluate the quality of the explanations. A common proxy is a decision tree as a highly interpretable model, but careful trade-offs need to be achieved between choosing a full interpretable proxy and a less interpretable proxy method that better represents the model behavior. The relation between interpretability and completeness helps also in better answering the question about the real need of explainability for ML models as we will discuss in the last section of this chapter.

Looking at the overall path we have followed so far, there is not an easy way to assess the level of explainability of a model quantitatively. And the main reason for this is that this level has to be assessed by a human with different knowledge of the domain and different needs from case to case (up to the point that the evaluation may not include humans at all as in the functionally grounded approach).

We set explainability as something more than interpretability, but at the same time we discussed how going toward explainability may mean increasing completeness (accurate description of the system to anticipate his behavior) but trading off interpretability (descriptions that are simple enough for a human with his specific domain knowledge).

The main takeaway of this digression is to use the factors we analyzed to properly set the priorities for XAI depending on the context. Looking again at the Fig. 2.4 below, we see that it is the same one we used to set the different types of evaluation but with an additional arrow on the left side: going up from functionally grounded to application-grounded evaluation usually means to increase cost and complexity. And we are now aware about the fact that increasing completeness of

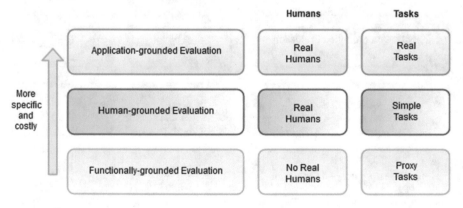

Fig. 2.4 Categories of explainability based on the role played by the human in the evaluation; arrow indicates increasing costs and complexity

explanations can reduce interpretability. With these coordinates in mind, we are in the position to make decisions about the right evaluation for XAI based on the goals and constraints.

2.2   How to Make Machine Learning Models Explainable

There is not a unique approach either to have a quantitative assessment of explainability or to define a taxonomy of XAI methods. Independently from the categories that we will use, it is important to have a real feeling of the factors that will orient the choices on the methods to make ML models explainable. Because of this, we will walk through a real case scenario that, albeit very simple, provides a tangible representation of these concepts. Suppose that you work in a marketing department and you are asked to provide a rough model to forecast the sales of smartphones depending of the age of the costumers. The rough data are shown in the Fig. 2.5 below.

Looking at the diagram, it is evident how the purchases are scattered in the plane without too much "regularity" going up and down. What we are saying here as "up and down" can be translated into a more formal knowledge of the learned function characteristics. Basically, we can distinguish three main behaviors of the learned function that give us indications about the level of explainability of the model:

- *Linear functions*: these are the most transparent class of Machine Learning models. Having linear functions means that every change in an input feature produces a change in the learned function at a defined rate and in one specific direction with a magnitude that can be read directly in an available coefficient of the model. Of course, all linear functions are also monotonic. Just to recall, a monotonic function is function that is increasing on its entire domain or decreasing on its entire domain.

Fig. 2.5 Purchases of smartphones depending on age of the customer

- *Nonlinear, monotonic functions*: in this case we don't have a single coefficient as direct evidence of the magnitude of the change in the output given a change in a feature input, but in any case, the change remains in one direction only for a given input feature variation.
- *Nonlinear, non-monotonic functions*: these are hard to interpret because the response function output changes in different directions and at different rates for changes in an input feature.

Let's get back now to our example but now trying to predict the sales. Figure 2.6 shows the case of a model trying to predict the number of purchases of smartphones based on the age of the consumer with a linear monotonic function.

The explanation is ready to go; the slope of the line tells us how much the number of purchases is expected to increase for a variation of one unit of age. Now look at the same problem in the Fig. 2.7 below but with the adoption of a nonlinear, non-monotonic function.

As you may easily realize, we lost the possibility of getting a single slope to explain the relation between the number of purchases and the age. The first linear model doesn't capture enough information of how purchases are related to age of the customers. This means that assuming that older customers are expected to buy more smartphones is not true across all the ages; there are intervals in which the trend changes. And to capture this trend, we need to add complexity to our ML model adopting a nonlinear non-monotonic function.

But the comparison between these two figures tells us another important thing: the linear model doesn't fit so well the observed data, so you are getting very good explanations at the price of losing accuracy on predictions.

And yes, for smart readers like you, this is a confirmation of the common belief we promised to demystify in Chap. 1: there is a trade-off between performance and explainability; getting more performance may reduce explainability, but keep

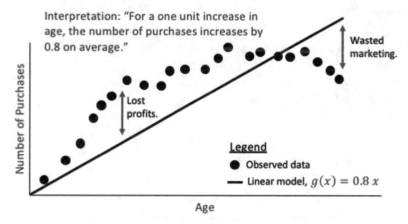

Fig. 2.6 A linear monotonic function gives simple, ready-to-go explanations with one global characteristic, the variation of purchases for one unit of age

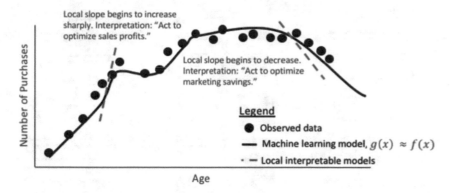

Fig. 2.7 With nonlinear non-monotonic function, we lose a global easily explainable model in favor of an accuracy improvement

trusting us and we will show how it is not always the case. We may guess a potential objection to our conclusion related to possible overfitting. We may argue that the second better performing model is overfitting the data and working on reducing overfitting would keep the performance and improve the explainability. At this stage, we say that this could be the case, but we are not yet able to detail this argument. Assuming no overfitting for the moment (the model here is just qualitative without any quantitative details), the idea behind this example is just to show how improving performance *in general* may challenge explainability, but we will see in Chap. 5 that the best practices for building better models (in terms of avoiding overfitting and bias among the others) may also help explainability and vice versa: explainability may guide ML model building.

Later in this chapter, we will look again at this figure to say something more about the dashed lines and what they mean in terms of explainability. Dashed lines represent local approximations of the model; it is like using a linear model and getting explanations that are valid only for a small interval to fit the local behavior of the more complex overall function. It is the time now to progress further toward taxonomy of the different XAI methods keeping in mind the ideas about the functions.

The first distinction is about *intrinsic explainability* and *post hoc explainability*. In the first case, the model is built already as "intrinsically" explainable (think about the linear model of the previous example above), while in the second case the explainability is achieved at a later time after the model creation (post hoc refers to this fact, that the explainability is achieved in a second phase with an existing and running model in place).

Figure 2.8 shows the categories that we will use to go through the different XAI techniques following the approach of Du et al. (2019).

All three schemes are about analyzing a Deep Neural Network (but the type of model is immaterial), and the first distinction is in fact between providing an intrinsic explanation and giving post hoc ones.

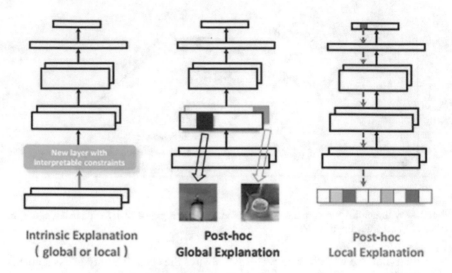

Fig. 2.8 Categories of different XAI techniques, Du et al. (2019)

The second level of distinction in the taxonomy that we envision is more about scope; whatever intrinsic or post hoc explanations, we may further split between global (user can understand how the model works across all the range of data) and local explanations (specific explanations are provided for individual predictions). Let's go to the next sections to better detail these general categories.

2.2.1 Intrinsic Explanations

There are two main classes of models that can be defined as intrinsically explainable (synonyms that can be used are white-box or interpretable models). In the first class, the subset of algorithms that produces the ML models can be directly interpreted. We take as representatives of this class the *linear regression*, *logistic regression*, and *decision tree models*. We will detail all of them in the next chapter, but it is important to start understanding why we call them intrinsically explainable.

Linear regression is used to model the dependence of a target (Y) from a set of features ($x_1 \ldots x_k$) with a linear relation:

$$Y = m_0 + m_1 x_1 + m_2 x_2 + \cdots + m_k x_k \tag{2.1}$$

Equation (2.1) Standard linear relation between a target Y and a set of features ($x_1 \ldots x_k$)

The obvious advantage in terms of explainability is that we have the weights ($m_1 \ldots m_k$) to compare to understand the relative importance of the different features.

This is a more general case of the example we did to forecast purchases. In that example, we have just one feature (age) to model the relation with purchases. Here we have more than one feature as purchases would depend not only on age but also on other factors (e.g., gender, salary). Whatever the situation, the weights directly show the importance of the feature in predicting the outcome.

Logistic regression is a variation on linear regression to handle classification problems. We will talk again about it in Chap. 3 with a real working case, but it is important to start familiarizing yourself with it.

Do you remember the example of dogs and wolves we described in Chap. 1? It is a classical classification problem that is not addressed well by linear regression. Assuming you have these two classes (dog vs wolf) and images to classify, if you heuristically try to use a linear regression model, it will fit and split the data between the two classes. But it just finds the best line (or hyperplane for more than two dimensions) to interpolate and split the set. That could be a problem.

In case of logistic regression, what we are searching for is something that gives us as output the probability that a specific item is a dog or a wolf and the probability by definition runs between 0 and 1. So in our case, we would set 0 for dogs and 1 for wolves. But using linear regression results with a fitting to the data that will produce numbers below 0 and above 1 which are not good for classification. So, you don't have a direct interpretation of the output as the probability for a given item. This is fixed in logistic regression but with a bit more complicated function than Eq. (2.1) above:

$$P(Y = 1) = 1/\left(1 + \exp\left(-\left(m_0 + m_1 x_1 + m_2 x_2 + \cdots + m_k x_k\right)\right)\right) \tag{2.2}$$

To deepen the concepts, Eq. (2.2) solves two theoretical problems in Machine Learning: it is the simplest equation that maximizes the likelihood function of probability theory, and it gives a fair robust convex loss function. This makes it resilient to noise and very easy to train. For such reasons, it has a very wide application even as part of complex models.

In the case of logistic regression, we don't have a direct mapping between the weights and the effect on the outcome (no linear relation), but it is still possible to make sense of them to produce explanations as we will see in the next chapter. Linear regression and logistic regression have in common the linear structure, such that if you draw in a bi-dimensional graph the samples you are trying to predict or classify, both linear regression and logistic regression split the different classes with a straight line. In this case we say that they both have linear decision boundaries.

The third main representative for intrinsic explainable models is *decision tree*. It is strongly different from logistic regression and linear regression. It doesn't have a simple linear decision boundary and can be used for both classification and

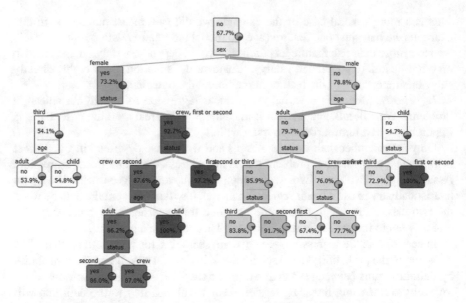

Fig. 2.9 A decision tree applied to the Titanic dataset. It is easy to explain how the model predicts the probabilities of survival

regression, but it is not based on a function to fit the data. It works by partitioning the information until the right subsets are identified. The splits are performed on putting cutoff values on the features, and the procedure can address also the case of nonlinear relations between features and outcomes but keeping a strong level of explainability. We can look at the picture below to have a first intuition on what is happening (Fig. 2.9).

The tree shows the probability of surviving the Titanic disaster based on some features as sex, cabin class, and age. There are two labels: "yes" is for survival and "no" for not survival. In each box you can see the percentual of that label. So in the first box, we can read "no" 67.7%, the proportion of people that have not survived, which also is impliedly saying that the complementary 32.3% has survived.

Traversing the tree, it is easy to get explanations of how the features are playing to determine the probability of survival. In the box below, we provide some further details about how decision tree works but consider that we will talk deeper again about decision trees in Chap. 3.

We want to be very clear here and avoid any misunderstanding with our readers. In this context we said just enough about linear regression, logistic regression, and decision tree to pass the idea of explainability as an "intrinsic" characteristic of the models without the need of any further technique to interpret them. Each of these three models will be explored in depth later. For now, it is important to get comfortable with the idea of "intrinsic explainability" to position it in the general taxonomy of the available methods.

To the reader who is acquainted with Machine Learning, we remember decision trees do their splitting to achieve the *maximum purity* in the target variable, which means they maximize how homogeneous the groups are.

If you have 10 bicycles and you are splitting by color, if you end with 10 red bicycles and 0 blue the group is 100% pure, while if you end with 5 red bicycles and 5 blue bicycles you achieved a 100% impure group. Both Gini index and entropy are quantitative and general measures of this idea of purity. *Gini index* is directly linked with the example we made, and we would have 0 for the purest case (10 red bicycles, 0 blue) and 0.5 for the worst case (5 red bicycles, 5 blue ones). *Entropy* is a bit more complex in terms of formula and shaped to have a measure of disorder in a group (that is another aspect of pureness), by the way the boundaries would be the same with 0 entropy with 10 red bicycles and 0 blue (that is also a very ordered group because you can achieve it in just one configuration and a full split on the target variable that is color) and entropy equals to 1 in the case of 5 red bicycles and 5 blues ones (disordered state in respect to color).

Beside the models that provide explainability by design, as mentioned, there is another class of intrinsic explainable models we'll call *tempered models*. In this case we may start from a model that is non-explainable but assuming we have the ability of modifying it (as with the specific case of a model we are building from scratch or we have access to its internals), we change it by adding interpretability constraints.

To get the idea of this, you may want to force the nonlinear model depicted in Fig. 2.7 (predicting purchases of smartphones depending from age) but force the relation to be monotonic, to guarantee that the direction of change in the outcome is always the same and simplify explainability (if a feature grows it always produces the same effect on the outcome). Yes, you may have already guessed that this would mean to get back to linear model of Fig. 2.6, and this is exactly what we would do to have a tempered model in this simple case. What is the risk here? The risk here is that forcing the constraints from outside the model may reduce the performance of the model albeit improving explainability as we already discussed from another perspective in Sect. 2.2 discussing Figs. 2.6 and 2.7.

(Yes, another time the myth seems to be confirmed about the trade-off between performance and explainability, but we will see an explicit example about XGBoost models in which adding explainability constraints will enhance the performance, so stay tuned and go with the flow. The spirit of this journey is to provide the information when needed not as general reference "all at once.")

2.2.2   *Post Hoc Explanations*

What if we are not in the case of "intrinsic explanations"? We can still rely on *post hoc explanations with two main variants: model-agnostic explanations and model-specific explanations*. As we said "post hoc" here means that the explanations are generated with an existing model already in place that is not always interpretable.

The techniques that belong to the *model-agnostic category* treat the model as a black box without accessing to the model's internal parameters. The strength of the agnostic methods is that they may be applied to any ML model to generate explanations. A typical example to get the idea before details is the "permutation feature importance."

In this method the relative importance of a feature compared to the others is evaluated by looking at how the predictions are impacted by a permutation of the values of that specific feature in the dataset. Indirectly we may build explanations leveraging the features that contribute more on building the output.

Only as note here to keep in mind and to be discussed in depth in Chap. 4, another promising attempt is to train an intrinsically explainable model from the output of the *black box* one. This approach is known as "knowledge distillation." We use the powerful and complex black-box model to search the complete dataset for the solutions and then use the results to guide a simpler and intrinsically explainable ML model to replicate the behavior but only on the narrowed-down solution space. Basically, the black-box model acts as a teacher for the explainable model to replicate the same results but with the possibility of providing explanations.

> We cannot stress enough the importance of model-agnostic explanations from a practical point of view. Suppose you are called to explain a model created by a data scientist, but you don't know anything of the methodology used to train the model, and you have to test the weakness of the model and explain its answers to someone else. In this case, you may use an agnostic method, maybe a local one like LIME or SHAP (wait for future chapters for details on these methods). Agnostic methods have the advantage of being extremely friendly to the user and easy to use so they can be used even from someone who doesn't know anything about Machine Learning or Computer Science. This is exactly the reason that is causing so much interest and investments on these methods in industry.

The other possibility is to rely on *model-specific explanations* that are built specifically for each model through the examination of the model's internal structure and parameters. Model-specific explanations are usually complex because they need to deal with the inner structure of the ML model. An example is to use back-propagation (widely known in ML) from the other way around: trace back the model from output to input following the gradients (direction of MAX change) to identify the feature that most contributes on building the output. Another example comes from decision trees.

There is a simple yet more powerful version of decision tree called Random Forest which is not explainable, but do you remember how decision tree partitions information? A decision tree partitions information using an indicator like Gini impurity or entropy (see the note above), so we can empirically say that sudden variation in this indicator corresponds to important decisions and a XAI system may weigh the importance of features leveraging the related variation of this indicator, that is, a measurement of information gain at each split until the final label is assigned.

Again, with the risk of being boring, we are not assuming that you may have a full understanding of how these methods work with these few lines, but it is important at this stage just to start distinguishing between the different approaches and getting an intuition of the main concepts.

At this point of our journey, we are still missing the practical skills to make these methods working (we will tackle that in Chap. 3), but we are able to distinguish between the two main families of XAI methods: the intrinsic explanations and the post hoc ones and see the reasons behind the choice. In one case, for intrinsic explanations the model can be interpreted as is, looking at the parameters, while for post hoc situations the model is not directly interpretable, and we need to work on it as an agnostic black box or by playing with model internals to get features' relative importance. Also, we placed these methods into the wider context of "man in the loop" paradigm to show how the humans that ask for explanations for XAI are also active players to enhance the overall performance of the ML system.

2.2.3 Global or Local Explainability

As we mentioned, the distinction between global and local explainability is in terms of scope. With global explanations, we are mostly interested in how the model works from a global point of view, the mechanism that generates the outputs. In the case of local explanations, we are searching for an explanation of a single prediction or outcome.

The idea here is to approximate the ML model locally near a given input with an intrinsic explainable model that can be directly interpreted (as above for linear regression, logistic regression, or decision tree). The local surrogate is used to make sense of specific predictions, but these explanations are valid only for the neighborhood of the specific input.

Let's look again at our model to predict purchases (Fig. 2.10):

The nonlinear, non-monotonic function fits very well with the data. It is hard to provide explanations, but the dashed lines suggest an approach: get local linear approximations of the function to provide explanations around some specific area. We won't have a single slope but different slopes in different regions to have interpretations of what's going on. The dashed lines are exactly the local approximation of the model we are talking about. As you may see, we need a different line for different inputs to get an easy explanation in terms of a linear regression model that can locally approximate the nonlinear, non-monotonic one.

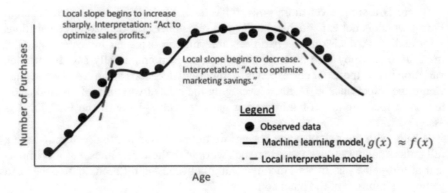

Fig. 2.10 Model to predict purchases. Dashed lines with different slopes can be applied to explain *locally* the predictions of the model

This example is also good to understand what we meant with the trade-off between interpretability and completeness. In case of linear regression, we had one single coefficient to explain the overall behavior of the number of purchases with the age. But we saw how the fit with the data was not so good. With a more complex function, we have better predictions, but we cannot provide explanations relying on a single parameter as before. Of course, we don't want to have function too complex to overfit data. To understand what's going on and anticipate the behavior of the system, we need to increase completeness of explanations. We rely on local approximations to interpret the results, so we produce different weights that have a validity limited to some set of data, losing a simple explanation for the overall. Providing explanations to humans in such a way could be difficult.

Let's try to be very clear about this point with our example. As we said, the scenario is that our marketing department has been asked to forecast the sales of a selected item depending on the age of the customers. We come back with a complex function that is very accurate in predicting the sales depending on age, but up to now this is just Machine Learning without XAI. XAI starts to play its role to explain to the people that assigned us the work "how" our model works in order to trust it. We have two options:

- *Achieve full interpretability*: we use the linear regression model instead of the complex function and say that the purchases are expected to increase with age.
- *Achieve completeness*: we cannot present the relation between purchases and age as global result but discuss the data with local linear surrogates. In some areas purchases increase with age; in others purchases decrease with increasing age. You may easily understand that this would make the scenario a bit more complicated to be explained and to be trusted. We are in the position of anticipating very well model's output with our local linear surrogates but losing some level of interpretability.

Another point needs to be considered to avoid oversimplification. In this first example, we supposed that the model relies on one feature, age. This is not like that

Fig. 2.11 A recap of explanations. We first divide models in intrinsically explainable and black-box models. We then add the scope of explanations: local or global ones. Finally we split the methodologies in agnostic and model-dependent

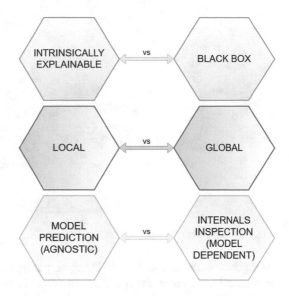

in the real world in which we would have used a lot of features to model the case (e.g., salary and sex, similar to the case of linear regression but with multiple variables). And producing explanations with multiple features becomes more complex because you lose the possibility of a quick look at the diagram to see what's going on as is the case with age. We used this simplification just to make more visible the relation between interpretability and completeness, but we will have the rest of the book to see how to deal in terms of XAI with huge number of features.

To recap with a different visual picture of what we have learned so far, we can have a look at Fig. 2.11.

The two big families of *intrinsic explanations* and *black-box methods* can be grouped in terms of what we are using to get explanations. In case of intrinsic explanations, we are mainly relying on "internals" because, as we saw, the parameters (or weights) may already provide the right level of explainability.

In case of black-box methods, we start from the model "predictions" and probe the model to understand behavior (how the predictions change), or we get local approximations valid for subset of the predictions; in alternative we use model-specific explanations *opening the box*, but also in this case we don't have "ready-to-go" explanations provided by the internals as in the case of intrinsic explanations.

Let's get back to our XAI flow presented in Chap. 1. As per Fig. 2.12, you might note that the classification of Du et al. (2019) used the term post hoc to put more emphasis on the fact that in this type of methods explainability is achieved at later time, after the ML model creation. The concepts do not change, but we prefer to use our flow (Fig. 2.6) in the rest of the book in which the first split is made between the ML models that are explained from the outside as black boxes (agnostic approach) and the ones that are explained looking at the internals (including the case of intrinsic explainable models). The other main difference is related to intrinsic explainable models: we consider them as a case of model-dependent approach

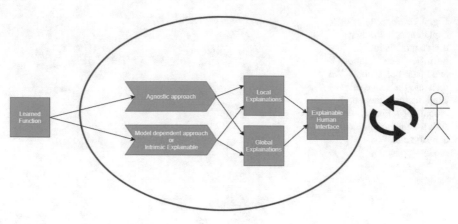

Fig. 2.12 XAI main flow

to produce explanations (while Du et al. put them in a dedicated category); albeit explainable, we rely on specific model internals to interpret them. We don't use the term post hoc, but it is important to keep it in mind as it is also present in XAI papers. We also think that it is better for you as reader to get used from the very beginning to different angles of looking at XAI considering the variety of views that you will be exposed in XAI literature as you dive deeper.

2.3 Properties of Explanations

We need a last step before starting hands-on work with Python code to do XAI starting in the next chapter. We want to group here the main terms that have a specific meaning as properties of explanations that will be used in following chapters. Usually we define terms when needed, but in this case, being the terms logically grouped we prefer to have all of them in one place to show how they are related.

As we said in the previous sections regarding XAI taxonomy, it is not currently possible to have a quantitative assessment of the XAI methods and generated explanations but just qualitative evaluations. The properties we are going to discuss in this section follow the same path. We want to be as precise as possible in characterizing explanations and XAI methods although we cannot assign a "score" for the different properties. We may envision a real case in which our "human in the loop" collaborating with ML model may provide a XAI report in which the methods and produced explanations are tagged with the properties we are going to discuss in the list below, following the work of Robnik-Šikonja and Bohanec (2018).

The first group is used to characterize the explanations and the methods from a global perspective (not the individual explanations):

Completeness: the accuracy of the description of the systems up to the possibility of anticipating the results of the model.

Expressive Power: this regards the language adopted for the explanations. As we are seeing, each explanation could be expressed in different ways, with different techniques, and for different "humans" (different needs, scope, and knowledge). There are different options for the explanations with different expressive power; they can be in the form of propositional logic (i.e., if-then-else), histograms, decision tress, or natural language to mention the main approaches.

Translucency: this describes how much the explanation is based on the investigation of the ML model internals. The two boundary cases we may take as an example are the interpretable models like linear regression, in which the weights are directly used to provide explanations, and then the methods in which inputs are changed to see the variation in output (common in agnostic methods) with zero translucency.

Portability: this assesses the span of Machine Learning models covered by the specific XAI method. Agnostic methods will have a high portability in general, while model-specific explanations will have the lowest portability.

Algorithmic complexity: this is related to the computation complexity of the methods to generate explanations. It is very important in terms of being a potential bottleneck to provide explanations in case of huge complexity.

The group below refers to properties of individual explanations (subset of the ones in Robnik-Šikonja and Bohanec 2018):

Accuracy: this is related to the usual definition of accuracy that comes out from ML but from a XAI point of view. In ML, accuracy is a performance metric defined as the number of correct predictions to the total number of input samples. For our purposes, accuracy is an indication of how well an explanation may predict unseen data. This is related to the argument of generating knowledge as an application of XAI. Explanation may be used for prediction instead of the ML model, and it should match at least the level of accuracy achieved by the ML system.

Consistency: this property describes the similarity between explanations generated from different models but trained on the same task. If the models generate similar predictions, the expectation is that related explanations are similar. We will see how consistency will be critical in the selection of XAI methodologies.

Stability: this property compares explanations between similar instances for a specific model. It differs from consistency which compares different models. If a small variation in a feature produces a huge change in the explanations (assuming that the same change has not produce a huge effect on the prediction), the explanations are not stable, and the explanation method has a high variance that is not good in terms of trust.

Comprehensibility: this is related to the arguments exposed in the "human in the loop" section of this chapter. The attempt is to have an idea of how well humans may understand the generated explanations.

2.4 Summary

- Evaluate explanations from a human point of view. Recognize the role of humans, and tailor explanations for specific audience.
- Distinguish between interpretability and completeness for explanations, and set the proper trade-off between the two depending on the main goal: achieve a detailed description of the system, or privilege easy explanations for the audience.
- Place XAI methods in a proper taxonomy:

  - Recognize intrinsic explanations vs post hoc explanations.
  - Use explanations in the right scope: global vs local.

- Locate XAI methods inside main XAI flow that will be used in the next chapters to address real case scenarios.
- Envision a report to have an assessment of explanations and adopted XAI methods, using the right properties.
- Learn how to think critically about XAI and challenge the real need of XAI to set the right level of explainability required.

In the next chapter, we will start hands-on work on interpretable models with specific examples leveraging Python. The goal will be to have practical cases to work on and provide explanations on the ML model predictions, relying on the properties of these models that are intrinsically explainable.

References

Doshi-Velez, F., & Kim, B. (2017). Towards a rigorous science of interpretable machine learning. *arXiv preprint arXiv:1702.08608.*

Du, M., Liu, N., & Hu, X. (2019). Techniques for interpretable machine learning. *Communications of the ACM, 63*(1), 68–77.

Gilpin, L. H., Bau, D., Yuan, B. Z., Bajwa, A., Specter, M., & Kagal, L. (2018). Explaining explanations: An overview of interpretability of machine learning. In *2018 IEEE 5th international conference on data science and advanced analytics (DSAA)* (pp. 80–89). IEEE.

Robnik-Šikonja, M., & Bohanec, M. (2018). Perturbation-based explanations of prediction models. In *Human and machine learning* (pp. 159–175). Cham: Springer.

Chapter 3
Intrinsic Explainable Models

"What I cannot create, I do not understand"
—Richard Feynman

This chapter covers:

- XAI methods for intrinsic explainable models
- Linear and logistic regression
- Decision tree
- K-Nearest neighbors (KNN)

The main objective of this chapter is to show how to provide explanations for intrinsic explainable models. As we said, for this category of ML models, XAI can be achieved by looking at the internals with the proper interpretations of the weights and parameters that build the model. We will make practical examples (using Python code) that will deal with the quality of wine, the survival properties in a *Titanic*-like disaster, and for the ML-addicted the evergreen categorization of *Iris* flowers.

Do you remember the XAI flow we envisioned in Chap. 1?

As you may see from the Fig. 3.1 and as already explained, we consider intrinsic explainable models in the same path of model-dependent approach for XAI: we can provide explanations because these models are not opaque, we can look at their internals, but each model is different from the other in terms of what to look at to provide explanations. Being intrinsic explainable, we won't have any issue on providing global (how the model works overall) or local explanations (give rationales for a single prediction).

We will focus on the concepts so that people may translate the same flows from Python to other programming languages or tools (e.g., R).

© The Author(s), under exclusive license to Springer Nature Switzerland AG 2021
L. Gianfagna, A. Di Cecco, *Explainable AI with Python*,
https://doi.org/10.1007/978-3-030-68640-6_3

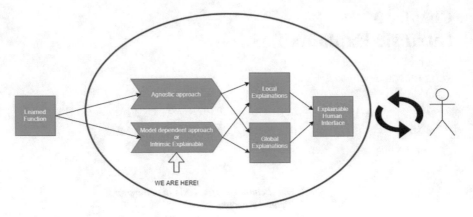

Fig. 3.1 XAI flow: intrinsic explainable models

3.1 Loss Function

Before going into details of our real-life scenarios, we need to revisit some concepts related to the loss function we already mentioned in Chap. 1. Do you remember the case we talked about before, about the risk associated to grant a loan to a customer? (Fig. 3.2)

As we said, if we model the error as a landscape, each choice of parameters makes a different model, and we prefer the model that minimizes the loss function. Every local minimum could be a good choice for the model. Each different model would generate a different set of interpretations. So, the loss function is fundamental from a purely ML perspective but also from a more specific XAI angle. (The loss function in the case of linear regression is also called *empirical risk.*)

A sufficiently powerful model such as neural networks can have many minima in the loss function, so the following situations produce a very complex loss function landscape:

- We have a terrible choice of features, some of which are irrelevant.
- Between samples there are strong outliers, samples that are very different from the majority of the samples.
- The problem itself is very difficult or badly posed (or, in fact, insolvable).

The connection we are making with XAI is that the regularization methods that are well known in Machine Learning and that are applied to handle such kind of complex situations also help in finding the most relevant features in a model.

Regularization makes the loss function smoother and provides indications about the relative importance of features in producing the output which helps provide explanations of the model.

Just to clarify: we already used the terms "feature importance" with different flavors (such as "relative feature importance"). There is no quantitative definition of this importance. The meaning is how much that feature contributes to building the output. Just as an example, we expect that the color of your favorite shoes won't contribute on predicting your health (low feature importance), while your weight might be an important feature in such a scenario.

As we know, we always express the training of a Machine Learning model via the minimization of a loss function. For example, we can express the loss as the mean of quadratic deviations from the expected value of the model as

$$\text{Loss}(w) = \frac{1}{2N} \sum_{i=1}^{N} \left(y_i - h(x_i; w) \right)^2 \tag{3.1}$$

Think of it as the total sum of errors you commit taking the outputs of a hypothesis function instead of the true values of the examples. A choice of parameters (weights) for the hypothesis function uniquely defines the model. If we plot the loss value of the model in the parameter space, we would expect something like (Fig. 3.3):

Finding the minimum of loss in the parameters is equivalent to choosing the best model, that is, satisfying the prescribed loss function.

The most used method for finding such a minimum is gradient descent (GD). In gradient descent, we simply recursively update the weights in the reverse direction of the gradient and proportionally at the gradient's magnitude (Fig. 3.4):

$$w_i = w_i - \eta \frac{\partial \text{Loss}}{\partial w_i} \tag{3.2}$$

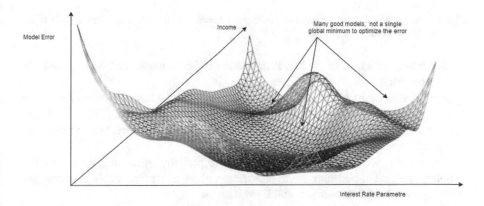

Fig. 3.2 An illustration of the error surface of Machine Learning model

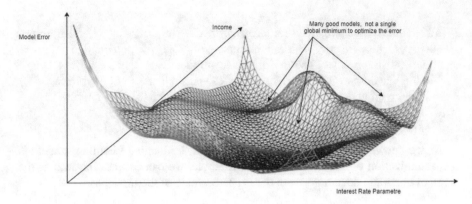

Fig. 3.3 Smooth loss function in a generic parameter space

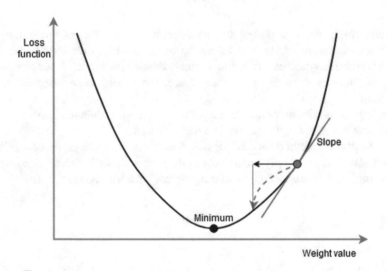

Fig. 3.4 In gradient descent we jump to the opposite direction to the gradient proportionally to the value of the gradient

For example, if locally the loss is increasing in space parameter, we decrease the parameter value.

From the courses on Machine Learning, we know GD is very sensitive to the initial position and the roughness of the loss function landscape. This affects the convergence of the training process, for example, we can be attracted and be stuck in a local minimum.

Technically, we can smooth the loss function partially by solving the roughness problem with *regularization*. So regularization gives us a more stable convergence of the gradient descent method, and it even accelerates it.

There are many forms of regularization especially in the domain of neural networks, but the most used ones are the *Tikhonov regularization* (*also called ridge or l_2 regularization*)

$$\text{Loss}_{\text{ridge}}\left(w\right) = \frac{1}{2N}\sum_{i=1}^{N}\left(y_i - h\left(x_i;w\right)\right)^2 + \lambda\sum_{i=1}^{M}w_j^2 \tag{3.3}$$

and the Lasso or l_1 regularization

$$\text{Loss}_{\text{lasso}}\left(w\right) = \frac{1}{2N}\sum_{i=1}^{N}\left(y_i - h\left(x_i;w\right)\right)^2 + \alpha\sum_{i=1}^{M}\left|w_j\right| \tag{3.4}$$

Adding a positive quadratic term in the parameters (or the sum of absolute values of weights) to the standard loss function transforms the loss into a smoother and more convex one. Moreover, the new positive term forces the weights to stay small, and this property gives less sensitivity to noise. In the absence of noise and outliers, ridge regularization usually gives more accurate results than Lasso regularization. It is even differentiable, but Lasso is more resilient to outliers because it gives less importance to greater deviations.

A remarkable property of Lasso regression is that it gives more weights to minor deviations. That helps us consider the true relevance of a feature and forces the weight relative to irrelevant features down to zero.

As an example, think of adding as a feature in a bank model for loans the zodiac sign of the loaner. With Tikhonov regularization you will have a small weight relative to the sign. With Lasso regularization that weight will be exactly zero.

So Lasso regularization will be one of our tools of choice to identifying the most relevant features in a model. We can even rank the features by feature importance.

We are now in the position of exploiting these concepts in the practical scenarios we will detail in the following.

3.2   Linear Regression

We start our journey through intrinsic explainable models with the application of linear regression to the problem of predicting the quality of red wine on a qualitative scale of 0–10, depending on some specific set of features. Suppose a wine producer heard about the "miracles" that Machine Learning may perform to improve business nowadays. The producer knows almost nothing about math, but he thinks that a good data scientist (i.e., you as the reader) may create insight into the chemical analysis of wine.

The idea of the wine producer is to leverage the results to increase the wine price and/or reposition the product on the market.

Here we are assuming that you as a data scientist decided to use a linear regression ML model to make a prediction about chemical analysis and wine quality. You want to provide explanations and interpretation of the results so to answer questions like: "Which characteristics of wine have more impact on quality?"

The tricky point will be that the answers to such questions are meant for the wine producer, so it won't be enough to show numbers. XAI must do its job of providing "human-understandable" explanations.

Remember that linear regression belongs to the category of intrinsic explainable models. This means that we are in a position to get interpretations directly from the model weights. But let's start by getting familiar with the wine quality data that we will use from UCI Machine Learning Repository (UCI 2009) (Table 3.1).

```
Wines.head()  #A


#The Wine-set as our dataset
```

As we see, we rely on the following 11 features to predict quality: "fixed acidity," "volatile acidity," "citric acid," "residual sugar," "chlorides," "free sulfur dioxide," "total sulfur dioxide," "density," "pH," "sulfates," and "alcohol."

It helps to take a first look at the general description of the feature statistics to see variation around max and min values.

```
Wines.describe()
```

A quick look at the table that is generated as output (we don't report it here because it is pretty large) shows the different scales for the various features besides the info about the min and max values and percentiles.

As we said, we will use linear regression to build our ML model and get predictions on wine quality. Let's repeat very quickly how linear regression would appear in two dimensions, as in the case in which quality would depend only on one feature (e.g., acidity) (Fig. 3.5).

The equation of the line that fits the data well is

$$Y = m_0 + m_1 x_1 \tag{3.5}$$

We use this two-dimensional simplification to have a visual explanation of the two weights m_0 and m_1:

- m_0 represents the quality value in case of acidity = 0 ($x|1 = 0$).
- m_1 represents the increase (decrease if negative) of quality for a unit increase of acidity.

This needs to be generalized to the case of multiple features as below (11 features in our case, $k = 1 \ldots 11$):

$$Y = m_0 + m_1 x_1 + m_2 x_2 + \cdots + m_k x_k \tag{3.6}$$

We can now write a few lines in Python to build the linear regression model for predictions. We will show below only the most meaningful lines (full code available with the book).

Table 3.1 The wine DataFrame

	Fixed acidity	Volatile acidity	Citric acid	Residual sugar	Chlorides	Free sulfur dioxide	Total sulfur dioxide	Density	pH	Sulfates	Alcohol	Quality
0	7.4	0.7	0	1.9	0.076	11	34	0.9978	3.51	0.56	9.4	5
1	7.8	0.88	0	2.6	0.098	25	67	0.9968	3.2	0.68	9.8	5
2	7.8	0.76	0.04	2.3	0.092	15	54	0.997	3.26	0.65	9.8	5
3	11.2	0.28	0.56	1.9	0.075	17	60	0.998	3.16	0.58	9.8	6
4	7.4	0.7	0	1.9	0.076	11	34	0.9978	3.51	0.56	9.4	5

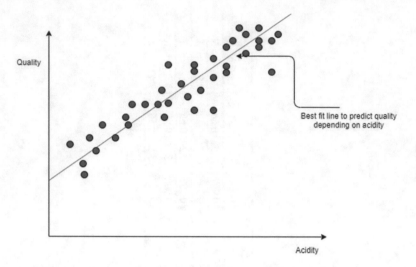

Fig. 3.5 Example of a linear regression in only two dimensions: acidity and the target (quality)

The code is based on the Scikit-learn free Machine Learning library for Python that everyone with some basic experience in Machine Learning knows and uses in everyday life.

You may see below the standard code to get linear regression on wine data.

```
df = pd.read_csv('winequality-red.csv')
X=df.iloc[:,:-1].values
Y=df.iloc[:,-1].values
x_train,x_test,y_train,y_test=train_test_split(X,Y,random_
state=3) #A

regressor = LinearRegression()
regressor.fit(x_train, y_train) #B

coefficients=pd.DataFrame(regressor.coef_,col_names)
coefficients.columns=['Coefficient']

#A usual splitting of data between train and test
#B Fitting to produce the coefficients
```

With the few lines above, we found the coefficients for the linear regression that provide us directly the explanations we are searching for in this specific case (Table 3.2).

```
print(coefficients.sort_values(by='Coefficient', ascending=False))
```

Table 3.2 Sorted coefficients
of the linear regression

Coefficient	
Sulfates	0.823543
Alcohol	0.294189
Fixed acidity	0.023246
Residual sugar	0.008099
Free sulfur dioxide	0.005519
Total sulfur dioxide	-0.003546
Citric acid	-0.141105
pH	-0.406550
Volatile acidity	-0.991400
Chlorides	-1.592219
Density	-6.047890

We see the ones that are negatively correlated with quality with a minus sign and the ones that positively impact the quality with a plus sign. The top three contributors are the negative values *density*, *chlorides*, and *volatile acidity*, while the top three positive ones are *sulfates*, *alcohol*, and *fixed acidity*. So, without relying on complex tools or artifacts, we have shown how we can make sense of a linear regression in terms of getting explanations about how the prediction about quality is produced.

With minimum effort and amount of code, you as the data scientist are already in the position of providing feedback to the wine producer. He wanted to know how to improve the quality of his wine. The direct answer is that he should work on the levels of sulfates and alcohol. Would that be enough? Would the wine producer trust our recommendation without any further explanation? We don't think so. We may try to bring him the considerations about the linear regression and the weights, but it would be hard relying only on these arguments to convince him about what to do to improve the quality.

The wine producer is not comfortable with mathematics and functions, so talking about the linear function that minimizes the loss function and the related coefficients would not help. We need to back up our results with some further artifacts to provide an effective explanation.

Do you remember that we talked about correlation in the previous chapters? Correlation is exactly what we need here to provide better explanations to our wine producer.

Correlation is a measure of the degree of the linear relation between two variables and can vary from −1 (full negative correlation, one variable's increase makes the other to decrease) to 1 (positive correlation, the two variables increase together). Every variable has obviously correlation = 1 with itself.

$$\frac{12(12-1)}{2} = 66$$

Table 3.3 Correlations with
our target "quality" of the
features of the model

Correlations with target	
Alcohol	0.476166
Volatile acidity	-0.390558
Sulfates	0.251397
Citric acid	0.226373
Total sulfur dioxide	-0.185112
Density	-0.174919
Chlorides	-0.128907
Fixed acidity	0.124052
pH	-0.057731
Free sulfur dioxide	-0.050554
Residual sugar	0.013732

Let's start looking at the table of correlation coefficients, of the 11 features with the output, below (Table 3.3):

```
correlations = df.corr()['quality'].drop('quality')
correlations.iloc[ (-correlations.abs()).argsort()]
```

Looking at the table values, we see how volatile acidity and alcohol are the features that are more correlated with quality (negatively for volatile acidity and positively for alcohol).

As we discussed about the difference between correlation and causation, this does not necessarily mean that alcohol and volatile acidity are "the main causes" of quality. For example, there could be a third unknown feature that control both alcohol and volatile acidity. But up to the boundaries of the current scenario with a linear model, volatile acidity and alcohol are the features that best "explain" the changes in quality.

One useful part about correlation is that it supports the possibility of a visual representation of what's going on through a visualization called a heatmap. Obviously, we can calculate the correlation of any two variables in the dataset. We have 11 features and 1 output (wine's quality); therefore, we can calculate 12 x 12 = 144 correlations, but due to the symmetry of correlations $\frac{12(12-1)}{2} = 66$ of them are unique.

With a few lines of code, we can generate the heatmap we mentioned (Fig. 3.6):

```
import seaborn as sns
 sns.heatmap(df.corr(),    annot=True,    linewidths=.5,    ax=ax,
cmap="twilight")
plt.show()
```

Looking at the picture, we may understand the name "heatmap" for the ones that are not used to it. It has to be read like a visual table in which each feature is

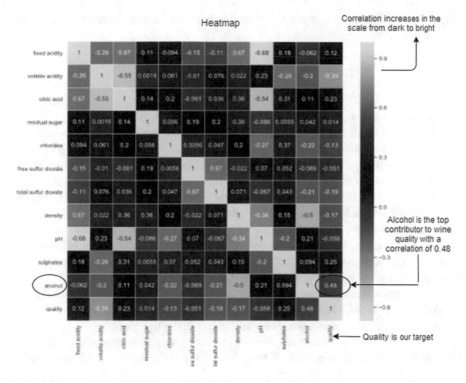

Fig. 3.6 Heatmap shows the correlation coefficients among the different features; focus is on correlation with our target: quality

correlated with another through a coefficient that corresponds to a different color with the scale shown on the right.

We see as expected that the main diagonal of the square is white (max correlation) because the diagonal shows the correlation of each feature with itself (= 1).

Apart from that, we may see the results that were numerically expressed in the previous table: we have the confirmation that *alcohol is the most important feature to improve quality.*

We have to stress that the present naïve linear model is not the state of the art for the solution for we have set aside three theoretical questions.

From a theoretical point of view, using only the weights gives a model that is scale-dependent. That is, if we measure density with a physical unit ten times larger, we will have a weight ten times smaller, so we can't directly compare the relative strengths between two weights, but we can do it with correlations which are adimensional.

This point is usually addressed by standardization, i.e., subtracting from the features their mean and scale dividing by the standard deviation of the features. The standardization gives us features with mean 0 and variance of 1.

We can easily see that the correlation of a coefficient with the target can have another interpretation. If we are in the case of a one-dimensional linear regression, we have the following formula between the m_1 coefficient of the regression and the correlation ρ between X and Y:

$$m_1 = \rho \frac{\sigma_Y}{\sigma_X} \tag{3.7}$$

Here σ_Y and σ_X are the standard deviations of Y and X.

So we can think the correlation as the regression coefficient in the special case $\sigma_X = \sigma_Y = 1$ of standardized quantities.

The second important question is the correlation between features: a heavy correlation between features gives a phenomenon called multicolinearity. Multicolinearity can yield solutions with error of weight that is extremely large, i.e., the weights are badly determined and possibly numerically unstable. Think of it, let's have two features x_1 and $x_2 = 2 * x_1$; here the correlation between features is 1. A general linear combination between features will be $w_1x_1 + w_2x_2 = x_1(w_1 + 2w_2)$. Using gradient descent we will exactly determine the weight $w_3 = (w_1 + 2w_2)$, but the relative contribution of w_1 and w_2 will be undetermined.

The usual method to overcome multicollinearity is to exclude features highly correlated or to *whiten* the features via principal component analysis. We don't resume these techniques for they are well found in the literature.

The third question is on badly determined weight even in absence of multicollinearity. A standard approach is to calculate the ratio of the coefficient and its uncertainty.

For a standard linear regression with one feature, we have the proportion

$$\frac{m}{\sigma_m} \sim \frac{m}{\left(\dfrac{\sigma_\varepsilon}{\sigma_X} \right)} \tag{3.8}$$

where σ_ε is the standard deviation of the regression and σ_X is the standard deviation of the feature. If we have a weight m smaller than its uncertainty σ_m, we can't determine even the sign of the weight, so low values of the ratio are undesirable. From the equation, we also see that the variance of a feature can't be too small.

Now we are roughly in the position of providing a better explanation to our wine producer: we may start showing the results of linear regression. In case he feels lost in the numbers, we may rely on the visual heatmap representation to support our explanation: our model produced a prediction that alcohol is strongly coupled with quality; changing alcohol level improves quality.

We are now ready to summarize the explanations we collected to answer the questions about the reasons of our predictions and recommendation on wine quality based on the outcome of the ML linear regression model.

Let's have a look at the Table 3.4 below:

The idea here is that in order to answer the question "Which are the main wine characteristics to work on to improve quality?," it is not enough to look at the weights and select those who have the biggest absolute value.

Table 3.4 Wine features' weights and correlation with quality

Feature	Weight	Correlation with quality
Alcohol	0.29	0.48
Sulfates	0.82	0.25
Density	−6.05	−0.17
Chlorides	−1.59	−0.13

We also need to look at the correlation with the target and to the variance of the feature. Density, for example, has the largest weight value but a very small correlation coefficient.

We have already seen that correlation is indeed the m of a linear regression when the target and the feature have been standardized. So the weight of the density *looks* big for the scale choice, but in effect it is small.

Also Eq. (3.5) reminds us that the uncertainty on weight could be even larger than the weight itself. We can resolve this problem restricting only to the more meaningful features.

The idea is simple but powerful. In Lasso regularization, we add a positive term in l_1 norm to force the weights associated with less significant features to be precisely zero.

If we gradually increase the Lasso constant, each feature will be zeroed one a time. The first to disappear will be the least important feature, and the last surviving feature will be the most important one. In this manner we can construct a more robust ranking of features by feature importance.

This procedure can be done only at training time. More useful techniques are done post hoc (using intrinsic or agnostic methods), i.e., after the model's training.

We standardize the features to unit variance and train a Lasso model with different alpha (Table 3.5):

```
x_train_scaled = preprocessing.StandardScaler().fit_transform
(x_train)
# scaling features

from sklearn import linear_model
regressor = linear_model.Lasso(alpha=0.045)
# selecting a Lasso regressor model

regressor.fit(x_train_scaled,y_train)
# training the Lasso regressor

coefficients=pd.DataFrame(regressor.coef_,col_names)
coefficients.columns=['Coefficient']
   print(coefficients.iloc[        (-coefficients.Coefficient.abs()).
argsort()])
```

Table 3.5 Coefficients after the feature selection with Lasso. The features have been standardized, and the α has been chosen to have six nonzero weights

Coefficient of lasso regression	
Alcohol	0.292478
Volatile acidity	-0.170318
Sulfates	0.079738
Total sulfur dioxide	-0.036544
Fixed acidity	0.020537
Chlorides	-0.002670
Citric acid	0.000000
Residual sugar	0.000000
Free sulfur dioxide	0.000000
Density	-0.000000
pH	-0.000000

For example with a choice of alpha = 0.045, we find and show what are the first six features by importance, and in fact "density" has been found as one of the less relevant features.

Do you remember the properties of explanations we explained in Chap. 2? Let's see how they fit with this real case scenario by looking at the Table 3.6 below.

3.3   Logistic Regression

In the previous section, we used linear regression to deal with wine quality and then getting explanation with XAI. In this section, we face a scenario of classification instead of prediction.

Suppose that a scholar in biology is using a ML classification system to distinguish between specific types of flowers. The focus here is not just on the classification but, given the classification with a certain accuracy, to provide explanations about the criteria that the ML model is adopting to assign the flower category. For this purpose, we will use the evergreen *Iris* flower dataset that is well known in ML community, shifting the focus from classification to explanations.

Let's state better our problem: we have a flower dataset containing instances of *Iris* species. We will build a ML logistic regression model to classify *Iris* flower instances as *Virginica, Setosa, or Versicolor* based on four features: pedal length, pedal height, sepal length, and sepal height. But our focus is not on the ML model that does the classification but on the explanations that we need to provide in this logistic regression case.

Assuming that the ML model does a good job on classifying the flowers, our goal is to provide methods to answer questions on how this classification is performed: which are the most important features among the four to split the plants into categories of *Iris Virginica, Setosa* or *Versicolor?*

From a XAI perspective, it is important to start from recalling why we adopt a logistic regression model for classification instead of an easier linear regression. Suppose for the moment that there is only one feature (the sepal length just for

Table 3.6 Properties of explanations

Property	Assessment
Completeness	Full completeness achieved without the need of trading-off with interpretability being an intrinsic explainable model
Expressive power	Correlation coefficients provide a direct interpretation of the linear regression weights
Translucency	High, we can look directly at the internals to provide explanations
Portability	Low, explanations rely specifically on linear regression machinery
Algorithmic complexity	Low, no need of complex methods to generate explanations
Comprehensibility	Good level of human-understandable explanations to build as much confidence as possible in the wine producer

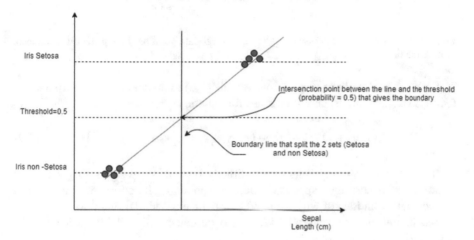

Fig. 3.7 Linear regression to classify *Iris* flower type based on sepal length only)

example) that controls if a flower in the dataset is an *Iris Setosa* or not, and suppose to have the situation depicted below (Fig. 3.7).

The threshold is the limit that we set to 0.5 as to get the probability that a specific *Iris* instance is *Setosa* or not. As we see from the figure, a linear regression would properly classify the two sets. Then suppose that you have a further data point like in the following graph (Fig. 3.8):

The additional point totally changes the regression line we use for predictions. In this new scenario, the previous set of flowers would be now categorized as non-*Setosa*, which is not the case. The way that one point changes the regression line shows how linear regression is not suitable for classification problems. This is the reason why we need to switch to logistic regression we already mentioned in the previous chapter.

Before jumping to the details of our flower classification problem, we need to get back to theory to understand what changes in terms of XAI, moving from linear regression to logistic regression. Mainly, we will lose the possibility of a straightforward interpretation of the model coefficients to provide explanations.

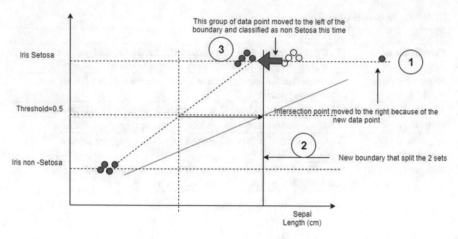

Fig. 3.8 Linear regression is broken by the additional data point on the top right; the numbers provide the flow

Let's get back to linear regression formula (3.6) to model target (Y) with a set of features ($x_1 \ldots x_k$); this is how it changes for logistic regression:

$$P(Y = 1) = 1 / \left(1 + \left(\exp - \left(m_0 + m_1 x_1 + m_2 x_2 + \cdots + m_k x_k\right)\right)\right) \qquad (3.9)$$

It is evident how we cannot rely directly on $m_1 \ldots m_k$ weights to get feature importance because they appear inside the exponential function. You may find in the box below an idea of why we need to use the exponential function, something to recall, but that is generally known to people familiar with basics in ML (Fig. 3.9).

> Let's review the need to use exponential functions. Basically, the problem we saw, where the addition of one data point may break the classification through linear regression, is fixed using the so-called logistic function $\sigma(t) = 1/(1 + (\exp - (t)))$ (Fig. 9).
>
> The effect that we obtain putting our linear regression formula instead of t is exactly the one we want: put a boundary on probabilities between 0 and 1 as it must be, and avoid the problem of the changing line with the additional data point.

So, what can we do to interpret the $m_1 \ldots m_k$ in terms of relative importance of features to produce explanations, as we did for linear regression case?

In Eq. (3.9), we are dealing with probabilities. In our specific scenario, we want to assign a probability for a flower classification. The first step is to tweak (Eq. 3.9) to have a probability ratio. We can understand the meaning of $P(Y = 1)$

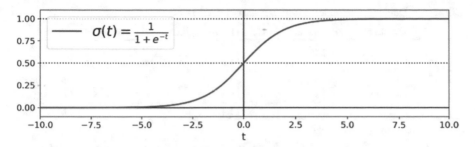

Fig. 3.9 Logistic function is used to fix the classification problems with linear regression

as the probability for a flower being an *Iris Setosa* and $1 - P(Y = 1)$ as the complementary one: the probability for the same flower does not belong in that category.

The first step is to take out the linear combination inside the exponential using the log function through some basic math:

$$\log \frac{P(Y=1)}{1-P(Y=1)} = \log \frac{P(Y=1)}{P(Y=0)} = m_0 + m_1 x_1 + m_2 x_2 + \cdots + m_k x_k \quad (3.10)$$

As we see the formula is back to linear but with log terms.

The term $\log \frac{P(Y=1)}{P(Y=0)}$ is named log-odds or *logit* where the odds mean the ratio of probability of the event and probability of no event.

This lets us express in a different way the relative weight of a feature compared to the others:

$$\frac{\text{odds}(x_k + 1)}{\text{odds}(x_k)} = \exp(m_k) \quad (3.11)$$

We skipped some intermediate steps to get formula (3.11) that are not fundamental for the main concept and may hide the result.

While in the case of linear regression m_k is directly the weight (the relative importance) of feature k, in the case of logistic regression we get something similar but a bit more complicated. With linear regression, the change of 1 unit in the feature k causes the target to change by the weight m_k. In logistic regression, the same change of 1 unit changes the odds by a multiplicative factor $\exp(m_k)$ all other things being equal. It may seem a bit abstract, but we will clearly need these concepts in our case of flower classification.

To anticipate a bit what will follows and to fix what we said with something more tangible, first assume m_k (just for example) to be the petal length. We may say that increasing it by one unit would enhance the probability for the flower instance of being an *Iris Setosa* by a factor $\exp(m_k)$. But to get to this point, we now need to have some Python code with real datasets and numbers. Let's have a first look at the dataset after the usual code to import libraries and load the *Iris* data (UCI 1988) (Table 3.7).

Table 3.7 Data sample extracted from *Iris* dataset (UCI 1988)

	Id	SepalLengthCm	SepalWidthCm	PetalLengthCm	PetalWidthCm	Species
0	1	5.1	3.5	1.4	0.2	*Iris Setosa*
1	2	4.9	3	1.4	0.2	*Iris Setosa*
2	3	4.7	3.2	1.3	0.2	*Iris Setosa*
3	4	4.6	3.1	1.5	0.2	*Iris Setosa*
4	5	5	3.6	1.4	0.2	*Iris Setosa*

```
X = iris.data[:, :2]  # we only take the first two features.
y = iris.target
df=pd.DataFrame(X, columns =
['Sepal_Length','Sepal_Width','Petal_Length','Petal_Width'])
df['species_id']=y
species_map={0:'Setosa',1:'Versicolor',2:'Virginica'}
df['species_names']=df['species_id'].map(species_map)
df.head()
```

Nothing special so far, just standard Python code to get the dataset; the sample shows some flower instances with the features and the species name. Next step is to train the model on the training dataset and then go for the classification on the test dataset as usual.

```
# Split the data into a train and a test set
perm = np.random.permutation(len(X))
f= df.loc[perm]
x_train, x_test = X[perm][30:], X[perm][:30]
y_train, y_test = y[perm][30:], y[perm][:30]

# Train the model
from sklearn.linear_model import LogisticRegression
log_reg = LogisticRegression()
log_reg.fit(x_train,y_train)
```

And after that we test the performance of the model:

```
# Test the model
predictions = log_reg.predict(x_test)
print(predictions)# printing predictions

print()# Printing new line


#Check precision, recall, f1-score
from  sklearn.metrics  import  classification_report,accuracy_
score
```

```
print( classification_report(y_test, predictions) )
print( accuracy_score(y_test, predictions))
```

The output is the following (Table 3.8):

['Versicolor' 'Setosa' 'Virginica' 'Versicolor' 'Versicolor' 'Setosa' 'Versicolor' 'Virginica' 'Versicolor' 'Versicolor' 'Virginica' 'Setosa' 'Setosa' 'Setosa' 'Setosa' 'Versicolor' 'Virginica' 'Versicolor' 'Versicolor' 'Virginica' 'Setosa' 'Virginica' 'Setosa" 'Virginica' 'Virginica' 'Virginica' 'Virginica' 'Setosa' 'Setosa']

The precision is the ratio `tp / (tp + fp)` where `tp` is the number of true positives and `fp` the number of false positives. We can think of precision as the ability of the classifier not to label as positive a sample that is negative.

The recall is the ratio `tp / (tp + fn)` where `fn` is the number of false negatives. We can think recall as the ability of the classifier to find all the positive samples.

The F-beta is simply the harmonic mean of recall and precision.

The support is the number of samples of y_test in each class.

All these concepts may be pretty clear to people that are familiar with ML. And this is the point where ML usually stops if we don't invoke XAI. Let's try to be very clear here: the ML model does a wonderful job on classifying *Iris* flowers, but from these metrics we don't have any evidence of the main features that are used to do this. How can our scholar in biology present the results? Which is more important for the classification, the sepal length or width? The petal length or width? These are the basic and natural questions that come from XAI and that we need to answer.

So let's do one step back and before going for the classification in three types, let's start from splitting *Setosa* from non-*Setosa* flowers to have a purely binary classification.

Do you remember Eq. (3.10) that we used to have something similar to linear regression to interpret coefficients?

Let's write the same equation but for our specific case in which we have our four features: sepal length, sepal width, petal length, and petal width for $m_1 \ldots m_4$. We saw how to use Eq. (3.11) to get the impact of each coefficient on the odds ratio; here we show how $m_0 + m_1 x_1 + m_2 x_2 + \cdots + m_k x_k$ can also be interpreted as the decision boundary for the classification.

Let's do a scatter plot of the flower dataset to understand this point better (Fig. 3.10).

Table 3.8 Scores of *Iris* classification using logistic regression

	Precision	Recall	F1-score	Support
Setosa	1.00	1.00	1.00	10
Versicolor	1.00	1.00	1.00	9
Virginica	1.00	1.00	1.00	11
Avg/total	1.00	1.00	1.00	30
1.0				

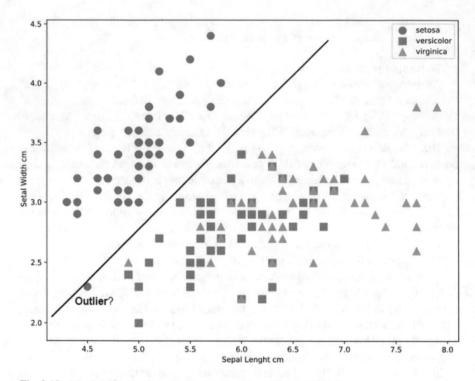

Fig. 3.10 *Iris* classification based on Sepal_Length and Sepal_Width features

```
marker_map = ['o', 's', '^']
unique = np.unique(df['species_id'])

for marker, val in zip(marker_map, unique):
    toUse = (df['species_id'] == val)
    plt.scatter(X[toUse,0], X[toUse,1], marker=marker,
cmap="twilight", label=species_map[val], s=100)

plt.xlabel('Sepal Lenght cm')
plt.ylabel('Setal Width cm')
plt.legend()
plt.show()
```

Setosa are the points marked as circles; the two sets are quite separated using and representing the scatter plot on features Sepal Length and Sepal Width.

Let's see what happens if we do the same but using Sepal Length and Petal Width (Fig. 3.11):

```
y = np.array(y)
marker_map = ['o', 's', 's']  # here we use same symbol for
versicolor and virginica
```

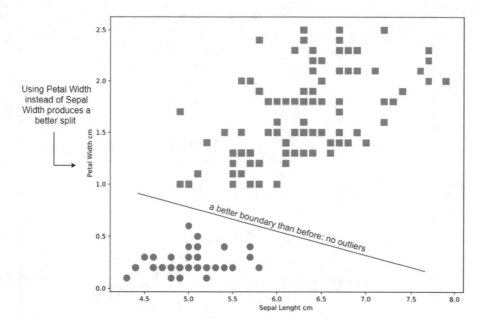

Fig. 3.11 *Iris* classification with Sepal Length and Petal Width instead of Sepal Width produced a better split

```
unique = np.unique(y)
for marker, val in zip(marker_map, unique):
    toUse = (y == val)
    plt.scatter(X[toUse,0], X[toUse,1], marker=marker,
cmap="twilight", s=100)
plt.xlabel('Sepal Lenght cm')
plt.ylabel('Petal Width cm')
plt.show()
```

The sets of *Iris Setosa* and non-Setosa are better separated. (We manually added the line that represents the boundary between the two groups.)

From this simple case on binary classification, we see how to identify the couple of features that are most important to classify *Iris Setosa* vs *Iris* non-*Setosa* flowers. Plotting the dataset on the plane of these features (Sepal Length and Petal Width), we have a clear linear boundary to delimit the two sets. But how to find the equation of this linear boundary to have a quantitative answer? It is easy to access the coefficients of the logistic regression model to get the linear boundary equation directly.

```
W, b = log_reg.coef_, log_reg.intercept_
W,b
```

```
Output: (array([[1.3983599 , 3.91315269]]),
array([-10.48150545]))
```

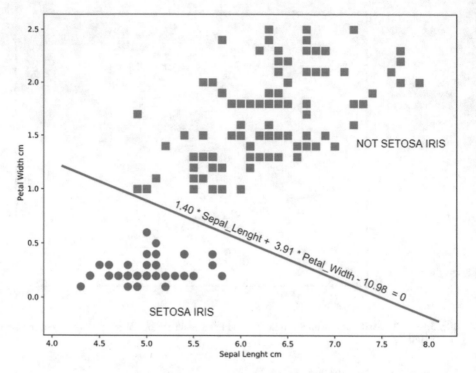

Fig. 3.12 *Iris* classification that shows the equation of the boundary line

What do these numbers mean? They are the coefficient of the linear boundary that separates the *Setosa* from non-*Setosa* flowers (Fig. 3.12):

$$\log \frac{P(Y = \text{Setosa})}{P(Y = \text{non-Setosa})} = \log \text{odds}(Y = \text{Setosa}) = m_0 + m_1 x_1 + m_2 x_2$$
$$= m_0 + m_1 (\text{Sepal Length}) + m_2 (\text{Petal Width}) \tag{3.12}$$

And the same coefficients are the ones that express the odds ratio:

$$\text{odds}(x_k + 1) / \text{odds} = \exp(m_k)(4)$$

For our specific case:

$$\frac{\text{odds}(x_1 + 1)}{\text{odds}(x_1)} = \exp(m_1)x_1 = \text{sepal length} \tag{3.13}$$

$$\frac{\text{odds}(x_2 + 1)}{\text{odds}(x_2)} = \exp(m_2)x_2 = \text{petal width} \tag{3.14}$$

Equations (3.13) and (3.14) express how the probability of *Iris* being *Setosa* or not changes for an increment of 1 cm in sepal length and petal width. And because of the scatter plots, we know that sepal length and petal length are the features that produce the best separation of the two sets. How would these arguments change in case of having three categories: *Setosa, Versicolor, and Virginica?* Not very much. We would do the same steps but with more coefficients and scatter plots to examine. The W and b of the previous equations would become:

```
W, b = log_reg.coef_, log_reg.intercept_
W,b
(array([[ 0.3711229 ,   1.409712   , -2.15210117, -0.95474179],
        [ 0.49400451, -1.58897112,  0.43717015, -1.11187838],
        [-1.55895271, -1.58893375, 2.39874554,  2.15556209]]),
 array([ 0.2478905 ,   0.86408083, -1.00411267]))
```

Having three categories would produce a 4*3 matrix of coefficients (four features, three categories) and three intercepts for the three categories. We would use these numbers to repeat what we did before for the binary classification and get the odds ratio to obtain the relative weights of the different features on explaining the classification (Table 3.9).

So what is the strategy for our scholar in biology to share his results from a XAI perspective?

Remember that one thing is just to build a ML model to classify flowers, another thing is to provide explanations of how this ML model is achieving the classification or, said in other terms, which are the most important features to recognize the *Iris* type.

Our scholar in biology would start explaining the results for the binary case. In particular, he could give a high-level overview of how odds are related to the probability of an *Iris* being a *Setosa* or not depending on the features. The assumption is that the audience could be comfortable with basics in mathematics. From a XAI perspective, the scholar could take the recommended further step to show how the coefficients are related to the liner boundaries that split the flower dataset into different types.

As shown in Fig. 3.10, we may point to sepal length and petal width as the features that are the most important ones for our classification problem. Then to further describe the explanations, the scholar may also explain how to obtain the odds ratio that expresses the feature weights, in the complete set of three flower species.

Let's repeat the assessment of explanations' properties that we already did for linear regression, using the Table 3.10 below:

Table 3.9 Features weigths

	Setosa	Versicolor	Virginica
Sepal length	1.44936119	1.61875962	0.21035554
Sepal width	4.09477593	0.20667587	0.20414071
Petal length	0.11623966	1.55210045	11.00944064
Petal width	0.38491152	0.33653155	8.63283183

Table 3.10 Properties of explanations

Property	Assessment
Completeness	Full completeness achieved without the need of trading-off with interpretability being an intrinsic explainable model
Expressive power	Less than linear regression case. Interpretation of coefficients is not so straightforward
Translucency	As any intrinsic explainable model, we can look at the internals. Weights are used to provide explanations but not so directly as in linear regression case
Portability	Method is not portable, specific for logistic regression
Algorithmic complexity	Low but not trivial as in linear regression case
Comprehensibility	Explanations are human understandable also for not technical people

3.4 Decision Trees

In this section we will be asked by an insurance company to explain our model of risk. Making this request more practical, we deal with the scenario of a marine insurance company, and we use a decision tree ML model to predict the survival probabilities. The insurance company is asked to provide explanations and criteria to back up the survival rates, and we will show how to get them.

We use a decision tree for, as we will see, they are the more logical choice in case of categorical tabular data. We download a reduced dataset of the famous disaster and use it to train and test a model that provides information on the fate of passengers on the Titanic, according to sex, age, and passenger's class (pclass). The target of the model is to predict survival as a Yes/No option. The insurance company has asked for explanations and criteria to bring up the survival rates, and we will show how to get them.

We have chosen this particularly easy example, well known in the ML field, to revise decision trees for those who already know Machine Learning basics and maybe want to deepen the concepts. We will focus on the aspects that we need from a XAI perspective.

Before going to the scenario and XAI, we need to review some concepts related to decision trees. Feel free to skip ahead if you feel comfortable with decision tree theory.

There are various implementations of decision trees. For simplicity, we will refer to the Classification and Regression Tree (CART) algorithm as implemented in Scikit-learn library. CART was introduced by Breiman in 1984 and is the first "universal" algorithm in the sense that it can accomplish both classification and regression tasks.

CART constructs a binary tree using some logic for splitting the initial dataset into branches so that the splits increasingly adapt data to the target labels.

The decision tree partitions the feature space into rectangles approximating the possible relations between features. It is similar to the case of a doctor that says: "If your weight is over 80 kg, you are at risk of diabetes" (Fig. 3.13).

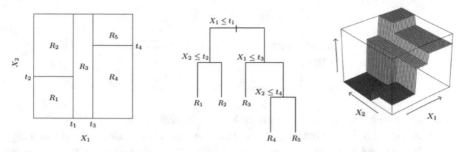

Fig. 3.13 Decision trees partition feature space approximating the true functional relation

So decision trees model human reasoning. With such a clear visual representation of the decision process, we can promptly answer counterfactual questions such as "What if?" and make explanations via contrastive arguments such as "your loan would be accepted if you were 5 years older" by simply looking at the representation.

Let's resume the theory behind decision trees.

We call impurity our primary indicator of how to do the splits.

For classification tasks like the dataset of Titanic, we can find in literature the different types of impurity:

- Gini impurity
- Shannon entropy
- Classification error

We call p_i the class proportion of occurrences in class i in respect to all the classes. The class proportion depends on the choice of the feature X_i and a pre-defined threshold value t_i for the split. In binary decision trees, every class is set of samples such that $X_i < t_i$ or $X_i > t_i$.

Now we can write the equations for the three impurity types:

Gini equation: $\text{Gini} = 1 - \sum_{i=1}^{C} (p_i)^2$

Shannon entropy: $\text{Entropy} = \sum_{i=1}^{C} -p_i \log_2 (p_i)$

Classification error: $\text{CE} = 1 - \max(p_i)$

Here C is the number of classes.

We know that for a perfectly classified item, impurity would be zero. As we already said in the previous chapter, a node is 100% impure when it is split evenly 50/50 and 100% pure when all node data belong to a single class.

So after we have selected the type of impurity to use in training a decision tree, we "simply" minimize the total impurity on each node finding the appropriate number of nodes and the corresponding couples of feature x_i and splitting threshold t_i.

We pick a loss function J as

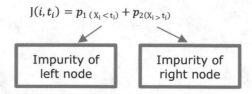

$$J(i, t_i) = p_{1 \, (X_i < t_i)} + p_{2(X_i > t_i)}$$

and minimize it finding the appropriate X_i and t_i in each node.

This choice of loss function is different from the loss function we have already seen. It is a *local* training loss function used for each node and not a global one. It defines the steps of the model training not the complete error on the training set.

We want to stress that the general problem of finding the best decision tree is too computationally expensive, in fact nonpolynomial in time (NP-Complete). So in practice, we employ a greedy strategy on the splits employing a top-down, greedy search to test each feature at every node of the tree.

For speed reasons, Gini impurity is the default choice of CART algorithm in Scikit-learn. As you can see in the following Fig. 3.14, the results don't change so much if you use it with respect to entropy.

In the case of regression, we build regression trees, and the impurity is simply variance. To train a regression tree is to find in each node the appropriate couples X_i and t_i that minimize prediction variance with respect to training samples. But we don't show further details. Now we will briefly talk about the advantages and disadvantages of decision trees with respect to linear regression and logistic regression we have already seen.

Decision trees can natively model nonlinear relations in the feature where, for example, linear regression cannot do it automatically. Also, linear regression does not consider the interaction between features such as how specific values

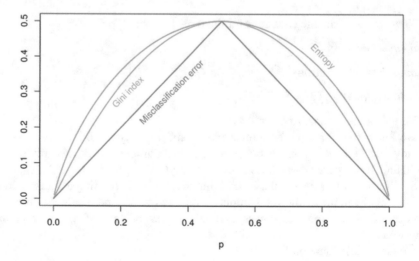

Fig. 3.14 Impurity as a function of class proportion p_i

of a feature constrain the variability of another feature. Decision trees can easily do that. We can add that using a greedy algorithm that gives fast computation time for the tree and is only needed to test enough attributes until all data is classified.

Decision tree models have a high capacity in describing (and memorizing) data as opposed to linear regression which has low capacity; in fact, they can have zero regression error on training data. But a high capacity comes with an increased risk of overfitting. As you know, overfitting means the model is so powerful that it "memorizes" data and noise alike but becomes incapable of predicting the outcome of some independent test samples. First variants of decision trees like ID3 addressed this problem following Occam's razor principle in the attempt to create the smallest possible decision tree. They selected the splits starting from those with higher purity gain.

But it was soon discovered that little gain in purity in some early node could give a substantial gain in the latter, so a criterion to make the DT simple would impact its performance. A more modern approach goes for deep trees, and after the training starts a pruning process in which the internal nodes of the tree not really needed to explain the total accuracy of the model are deleted, simplifying the tree. A simpler model is also more robust to outliers.

Decision trees give to us understandable prediction rules that are created from the training data in a directly interpretable manner.

The limit of such intrinsic interpretability lies in the depth of trees. In fact, the same feature can be reused in a different part of the same tree, and as we have seen little variation in the purity of a node in early splits can give us large effects later. So for interpreting feature importance in decision trees, we have to track for each feature the TOTAL PURITY VARIATION by simply adding all variations of that feature in every apparition of that feature.

With these concepts in mind, we can go back to the scenario we presented at the beginning of this section: an insurance company needs to provide explanations about its ML model that predicts survival rates for marine accidents. We use a model based on the famous Titanic dataset and learn how to explain its forecasts. Having categorical data, we decide to use a decision tree model. Using Scikit-learn we will train the model to explain its predictions and, as support, will calculate Permutation Importance as a measure of feature importance.

Behind the Permutation Importance methods is the following: assuming to have a trained model, we shuffle the values in a specific column and do the predictions again using the shuffled dataset. The predictions are expected to worsen because of the shuffling (dataset has been hacked!). We repeat the shuffling for each column (one column per time) and see which columns have more impact on the prediction (deteriorating more the performance). The columns that cause more deterioration are the most important features. Said in other terms, if shuffling the values of a feature screws up the model, then the ML model is heavily relying on this feature to do the job. That feature is expected to be very important.

Feature importance is not specific of decision trees. It is an agnostic method that works not knowing anything about the model's inner workings.

We stress that Permutation Importance in its simplicity is a post hoc procedure. We check the model workings against some test dataset AFTER the training process of the model. We can calculate Permutation Importance with just a line of code in Eli5. We have introduced a post hoc technique at this early stage of the book for its simplicity and to compare its results with the other methods we have already introduced.

We can now go to our code to work on our scenario; as usual let's start by having a look at the dataset (Waskom 2014).

```
#Load the data from Seaborn library
titanic = sns.load_dataset('titanic')

#Print the first 10 rows of data
titanic.head(10)
```

	survived	pclass	sex	age
0	0	3	male	22.0
1	1	1	female	38.0
2	1	3	female	26.0
3	1	1	female	35.0
4	0	3	male	35.0
5	0	3	male	NaN
6	0	1	male	54.0
7	0	3	male	2.0
8	1	3	female	27.0
9	1	2	female	14.0

We select convert the categorical data in dummy numerical ones:

```
from sklearn.preprocessing import LabelEncoder
labelencoder = LabelEncoder()
##Encode sex column
titanic.iloc[:,2]= labelencoder.fit_transform(titanic.
iloc[:,2].values)
```

And split features and target columns:

```
#Split the data into independent 'X' and dependent 'Y'
variables
X_train = titanic.iloc[:, 1:4].values
Y_train = titanic.iloc[:, 0].values
```

We train the model

```
from sklearn.tree import DecisionTreeClassifier
tree = DecisionTreeClassifier(max_depth=3)
tree.fit(X_train, Y_train)

#output of training
DecisionTreeClassifier(class_weight=None, criterion='gini',
max_depth=3,
          max_features=None, max_leaf_nodes=None,
          min_impurity_decrease=0.0, min_impurity_split=None,
          min_samples_leaf=1, min_samples_split=2,
          min_weight_fraction_leaf=0.0, presort=False,
          random_state=None, splitter='best')
```

and with a little of effort graph, the corresponding learned Tree

```
from IPython.display import Image
from sklearn.externals.six import StringIO
from sklearn.tree import export_graphviz
import pydot

dot_data = StringIO()
export_graphviz(tree, out_file=dot_data,feature_names=features
,filled=True,rounded=True)

graph = pydot.graph_from_dot_data(dot_data.getvalue())
Image(graph[0].create_png())
```

We can now explain entirely the ratio behind each prediction of the model (Fig. 3.15). This is exactly what we demand from an intrinsically explainable model, but which features are more relevant? This is the fundamental question from a XAI perspective. We are asked not just to do predictions but also to provide explanations.

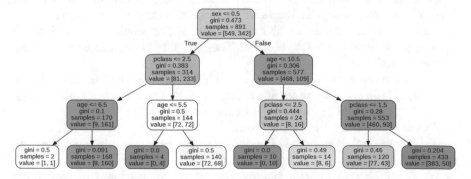

Fig. 3.15 Decision tree to predict survival rates based on the different features

We have two ways of assessing feature importance in DT. One is ranking the features by the most relevant decrease in Gini index, but as we already said such a method is neither robust or consistent, so we merely calculate Permutation Importance in this case (for the reasons we explained above).

So far, we have skipped stating the obvious about Scikit-learn installation and main usage, but it is important to note that Scikit-learn provides dependencies for the ELI5 package. ELI5 stands for "explain it like 5 (years old)" and is a well-established explainability library. As with any common ML Python package, it is enough to import ELI5 to access its features:

```
import eli5   #A
eli5.show_weights() #B


#A import eli5 package
#B Use show_weights() API to display classifier weigths
```

In the lines above, we import ELI5 and call one of its APIs just to show the syntax. We can now calculate Permutation Importance with just a line of code in Eli5.

```
import eli5
from eli5.sklearn import PermutationImportance
perm = PermutationImportance(tree, random_state=1).fit(X_
train, Y_train)
eli5.show_weights(perm)
```

Weight	Feature
0.1481 ± 0.0167	Sex
0.1003 ± 0.0152	Age
0.0301 ± 0.0105	Pclass

So sex and age are much more critical than passengers' class.

Look at the decision tree graph. Each node (rectangle) contains both survived and not-survived passengers. The *value* array contains the number of survived and not-survived passengers. So in the first rectangle of 891 people, the majority, 549 passengers, will not survive the accident. The first split puts women on the left and men on the right.

Now in the two subsets, the women subset has a majority who has survived. The second split is on age, and we see that in the male subset on the right splitting by age tells us that male children have a higher chance to survive.

We can argue that, greedily, the features we use first are more relevant, but it is not valid in general. Permutation Importance reassures us that this is precisely the case in this example.

We close the section with summarizing some disadvantages of pure decision trees.

Table 3.11 Properties of explanations

Property	Assessment
Completeness	Full completeness achieved without the need of trading-off with interpretability being an intrinsic explainable model
Expressive power	High expressive power; in fact DTs mimic to some extent human reasoning
Translucency	Intrinsic explainable easy to guess results
Portability	In fact many models are derived from decision trees such as Random Forest and boosted trees so DT results can be incorporated in such models
Algorithmic complexity	Decision trees are NP-complete, but we resort to heuristic for fast evaluation
Comprehensibility	Easy explanations to humans

The regions in the feature space will always be rectangular (we use only a feature at a time), and the transition from a region to another will not be smooth. In fact, decision trees struggle to describe even linear relations between features. A variation of DT called MARS is an attempt to add smoothness and natively nonlinear and nonlinear relations between features retaining intrinsic explainability.

As for properties (Table 3.11):

3.5 K-Nearest Neighbors (KNN)

We now return to the task of wine quality prediction but using K-nearest neighbors (KNN), another useful and intrinsically explainable methodology.

KNN was introduced in an unpublished report by the US Air Force School of Aviation Medicine in 1951 by Fix and Hodges, and it is one of the more established Machine Learning models of all Artificial Intelligence. Remember that we want to understand which features are the most important ones to increase the quality, and KNN will provide deeper insight on the explanations.

A KNN is easily explainable both using counterfactual examples that give visual explanations like "What if wine's acidity would increase?" and with a little modification even contrastive examples. With contrastive explanations, we create descriptions based on the missing abnormalities.

We can look to classification results with similar features. For each near sample with a different classification, we look for abnormalities in the corresponding features.

The core idea of KNN is to train the model merely memorizing all the samples and making predictions by an average or a majority voting process involving the results of some memorized examples (in fact, k of them) that have features most similar to the features of the item we want to predict.

In the example of the Titanic dataset, a passenger is predicted to survive if at least four out of seven passengers with similar features (same age, same boarding class, and same sex) have survived.

Technically, KNN differs from other learning algorithms both in training complexity, which is merely O(1), and in inference complexity that is much slower than other methods having to sort the samples continuously by the nearness. Without using some heuristic inference, complexity is O(N^2).

So let's go back to the wine producer who asked us how to improve the quality of wine and the reasons behind such answers. Just for visualization, we pick a large k to reduce the noise that affects the data, and after the splitting of data in a training set and a test set, we use only two features for the model.

```
# Importing the dataset
dataset = pd.read_csv('wine_data.csv')
X = dataset.iloc[:, 1:13].values
y = dataset.iloc[:, 0].values


# Splitting the dataset into the Training set and Test set
from sklearn.model_selection import train_test_split
X_train, X_test, y_train, y_test = train_test_split(X, y,
test_size = 0.10)


# Fitting KNN to the Training set
from sklearn.neighbors import KNeighborsClassifier

classifier=KNeighborsClassifier(n_neighbors=15, metric=
"euclidean")
trained_model=classifier.fit(X_train[:,0:2],y_train)
```

We draw the boundaries using a mesh grid. For each node in the grid, the model predicts the corresponding class

```
X=X_train
h=0.05
x_min, x_max = X[:, 0].min() - 1, X[:, 0].max() + 1
y_min, y_max = X[:, 1].min() - 1, X[:, 1].max() + 1
xx, yy = np.meshgrid(np.arange(x_min, x_max, h),
                     np.arange(y_min, y_max, h))
Z = trained_model.predict(np.c_[xx.ravel(), yy.ravel()])
kk=np.c_[xx.ravel(), yy.ravel()]

# Put the result into a color plot
Z = Z.reshape(xx.shape)
plt.figure(figsize=(14, 8))
plt.pcolormesh(xx, yy, Z)
plt.scatter(X[:, 0], X[:, 1], c=y_train)
plt.title("Wine KNN classification (k = 15)")
plt.show()
```

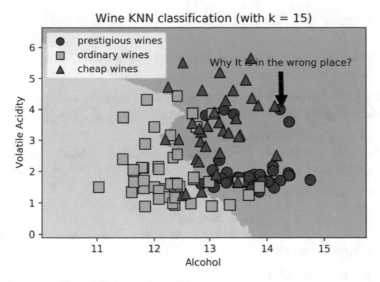

Fig. 3.16 Wine quality classification with KNN

Table 3.12 Explanations' properties

Property	Assessment
Completeness	Full completeness achieved without the need of trading-off with interpretability being an intrinsic explainable model
Expressive power	High expressive power in terms of counterfactual and contrastive explanations
Translucency	Intrinsic explainable easy to guess results
Portability	KNN has a unique class in its own not portable
Algorithmic complexity	Simple training, complex inference step
Comprehensibility	Easy explanations to humans

Circles correspond to top-quality wines and triangles to poor-quality ones (Fig. 3.16). The *x* axis represents alcohol content and the y axis volatile acidity.

Now you can clearly visualize using the figure which examples in the training set have a similar composition to a new wine for which you want to predict the quality.

And you can even see which wines are "abnormal" in the sense that they are members of other classes of wine quality and they follow the regularities of a different class.

So you can give counterfactual explanations ideally changing the position of wine in the picture. For example, "If you decrease acidity you will go in the red zone of more prestigious wines."

And you can also give contrastive explanations pointing at the regular structure of blue points in the blue "poor wines" zone. For example, "All cheap wines tend to have a volatile acidity greater than 2."

We close with the usual properties (Table 3.12):

3.6   Summary

We have seen how intrinsic interpretable models can be interpreted and how to produce "human-understandable" explanations:

- Use l_1 regularization for XAI to get feature importance.
- Produce explanations for linear regression models:
  - Use weights to rank feature importance.
  - Interpret correlation coefficients to produce human understandable explanations.
- Provide explanations for logistic regression models:
  - Use log-odds to provide explanations.
  - Match the logistic regression coefficients with decision boundaries to enrich the explanations of the results.
- Interpret decision tree models:
  - Extract decision tree rules for explanations.
  - Use Permutation Importance technique to provide feature importance.
  - Mitigate the limitations of decision tree models.
- Provide counterfactual explanations using KNN models.

In the next chapter, we will start our journey through model-agnostic methods for XAI. The main difference is that we will lose the possibility of "easy" explanations as for the case of intrinsic explainable models, but we will learn how to get explanations through powerful methods that can be applied "agnostically" to different ML models.

References

UCI. (1988). *Iris data set*. Available at http://archive.ics.uci.edu/ml/datasets/Iris/.
UCI. (2009). *Wine quality data set*. Available at https://archive.ics.uci.edu/ml/datasets/wine+quality.
Waskom, M. (2014). *Seaborn dataset*. Available at https://github.com/mwaskom/seaborn-data/blob/master/titanic.csv.

Chapter 4
Model-Agnostic Methods for XAI

"Then why do you want to know?"
"Because learning does not consist only of knowing what we must or we can do, but also of knowing what we could do and perhaps should not do."
—Umberto Eco, The Name of the Rose

This chapter covers:

- Permutation Importance
- Partial Dependence Plot
- Shapley Additive exPlanations (SHAP)
- Shapley Values Theory

In this chapter, we start our journey through XAI model-agnostic methods that are, as we said, potent techniques to produce explanations without relying on ML model internals that are "opaque."

The main strength of model-agnostic methods is that they can be applied to whatever ML model, including intrinsic explainable models. The ML model is considered as a black box, and these methods provide explanations without any prerequisite knowledge of ML model internals. We want to be very clear on this last statement: as you may remember, in Chap. 3 we used Permutation Importance to produce explanations on the Titanic decision tree.

In that case, we relied on the "intrinsic explainability" of the decision tree, but we saw how Permutation Importance produced enhanced interpretability. Here we start again from Permutation Importance methods but applied to a case in which we don't have the possibility of producing intrinsic explanations (Fig. 4.1).

Back to our flow, as shown after, we are going through the path of an agnostic approach, and we will distinguish, for each method, the local or global scope of the provided explanations.

© The Author(s), under exclusive license to Springer Nature Switzerland AG 2021
L. Gianfagna, A. Di Cecco, *Explainable AI with Python*,
https://doi.org/10.1007/978-3-030-68640-6_4

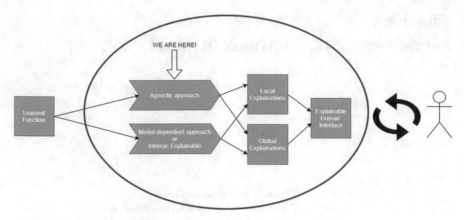

Fig. 4.1   XAI flow: agnostic approach

We will use two main real-case scenarios to explain how these methods work. The first one is based on a scenario presented in XAI Kaggle learning section that we recommend (Becker 2020): basically, it deals with a Machine Learning division of a company that is in betting business for sporting events.

You will learn how to answer "What" and "How" questions on the scenario using XAI agnostic methods. Permutation Importance will allow the identification of the most important features while Partial Dependence Plot will provide details on "How" the features are impacting the predictions. Moreover, you will be able to use SHAP to produce explanations on specific instances instead of just global explanations.

Having in mind how to use these methods, then we will provide theoretical foundations of SHAP before switching to the scenario of a taxi cab company to provide to customers explanations on the fares (predicted by a boosted tree ML model) they will pay in real time.

We will use the cab scenario to revisit SHAP in comparison with LIME and see how it performs with boosted trees. Also you will learn SHAP's limitations and how to tackle them.

4.1   Global Explanations: Permutation Importance and Partial Dependence Plot

We are in the Machine Learning division of a company that is in betting business for sporting events. We rely on a complex Machine Learning model that takes as input team statistics, and it predicts, as output, whether or not a team will have the "Player of the Match" prize (Becker 2020). This prediction is then passed to bookmakers in real time. Our stakeholders are concerned about how the ML model works and which are the most important criteria used to predict the team having or not "Player of the Match" award. As XAI experts, the first question we are asked to answer is: *What are the more significant features of your model?*

4.1.1   Ranking Features by Permutation Importance

The Permutation Importance method tries to answer this specific question. We recall what is already explained in Chap. 3: we assume to have an "opaque" trained model to interpret and understand the relative importance of the features leveraged by the model to make predictions.

What we do is to shuffle the values in a column (feature) and make the prediction again but with the shuffled values.

The expectation is that the error associated with the prediction will increase depending on the importance of the shuffled feature: the more important is the specific feature, the more the predictions will worsen because of the shuffling.

Assuming that the model is not relying on a feature to make the predictions, the shuffling of the values of that feature won't impact the performance.

So let's go to our scenario with the first snip of code here:

```
import numpy as np
import pandas as pd
from sklearn.model_selection import train_test_split
from sklearn.ensemble import RandomForestClassifier

data = pd.read_csv('../input/fifa-2018-match-statistics/FIFA 2018
Statistics.csv')
y = (data['Player of the Match'] == "Yes")  # Convert from string
"Yes"/"No" to binary
feature_names = [i for i in data.columns if data[i].dtype in [np.
int64]]
X = data[feature_names]
train_X,   val_X,   train_y,   val_y   =   train_test_split(X,   y,
random_state=1)
my_model = RandomForestClassifier(n_estimators=100,
                                  random_state=0).fit(train_X, train_y)
```

We use the statistics available from FIFA, after importing the file and cleaning the data; we just do the usual ML split between training and test data before building our model that, in this case, is a Random Forest (Becker 2020).

Let's have a look at the data to make sense of what's going on. The Table 4.1 is extracted from the original CSV file.

As you may see, we have the teams, the date, a lot of features, and our target feature that is Player of The Match (Y/N) saying if Team had the Player of the Match prize or no for that specific match against the "Opponent."

Table 4.1 FIFA 2018 matches, target feature is Man of the Match column

Date	Team	Opponent	Goal scored	Ball possession %	Attempts	On-target	Off-target	Man of the match
14-06-2018	Russia	Saudi Arabia	5	40	13	7	3	**Yes**
14-06-2018	Saudi Arabia	Russia	0	60	6	0	3	**No**
15-06-2018	Egypt	Uruguay	0	43	8	3	3	**No**
15-06-2018	Uruguay	Egypt	1	57	14	4	6	**Yes**
15-06-2018	Morocco	Iran	0	64	13	3	6	**No**
15-06-2018	Iran	Morocco	1	36	8	2	5	**Yes**

Table 4.2 Statistics after a first data manipulation needed to build the ML model

	Goal scored	Ball possession %	Attempts	On-target	Off-target	Blocked	Corners	Offsides	Free kicks
0	5	40	13	7	3	3	6	3	11
1	0	60	6	0	3	3	2	1	25
2	0	43	8	3	3	2	0	1	7
3	1	57	14	4	6	4	5	1	13
4	0	64	13	3	6	4	5	0	14
5	1	36	8	2	5	1	2	0	22

After our data manipulation aimed at building our Random Forest model, the data look like this (Table 4.2):

```
X.head()
```

The relevant thing to note is that we just kept the numerical data, transformed the Player of the Match (Y/N) in 1 or 0 values; the first column left to "Goal Scored" is a numeric id for the match.

As we said, we are using Random Forest to do "Player of The Match" predictions, but we are not interested in the ML details of Random Forest; we are focused on how to generate explanations around the predictions provided by our ML model. Random Forest is built as a huge ensemble of decision trees; each decision tree contributes to the final prediction that will result as the "most voted" one, a kind of wisdom of the crowd. It performs better than a decision tree at the cost of losing the intrinsic interpretability of a single decision tree. We will answer the question "What are the most important features of your model?" with Permutation Importance method, and the same flow can be adopted for any other opaque model, different from Random Forest.

Out[14]:

Weight	Feature
0.1750 ± 0.0848	Goal Scored
0.0500 ± 0.0637	Distance Covered (Kms)
0.0437 ± 0.0637	Yellow Card
0.0187 ± 0.0500	Off-Target
0.0187 ± 0.0637	Free Kicks
0.0187 ± 0.0637	Fouls Committed
0.0125 ± 0.0637	Pass Accuracy %
0.0125 ± 0.0306	Blocked
0.0063 ± 0.0612	Saves
0.0063 ± 0.0250	Ball Possession %
0 ± 0.0000	Red
0 ± 0.0000	Yellow & Red
0.0000 ± 0.0559	On-Target
-0.0063 ± 0.0729	Offsides
-0.0063 ± 0.0919	Corners
-0.0063 ± 0.0250	Goals in PSO
-0.0187 ± 0.0306	Attempts
-0.0500 ± 0.0637	Passes

Fig. 4.2 Permutation importance output. Every weight is shown with its uncertainty (Becker 2020)

```
import eli5  #A
from eli5.sklearn import PermutationImportance
perm = PermutationImportance(my_model, random_state=1).fit(val_X,
val_y) #B

eli5.show_weights(perm, feature_names = val_X.columns.tolist())
#C
```

#A *Here we Import eli5 library*
#B *Train The permutation importance model on the validation set*
#C *Show the feature importance*

Few lines of code are enough to generate our explanations with Permutation Importance (Fig. 4.2):

Let's explain the table. The features are ranked by their relative importance, so the first and most important result is that we may directly answer our question: "What are the most important features of your model?" – Goal Scored is the most important feature that our Random Forest ML model uses to predict if the team will have or not the Player of The Match prize.

The importances are calculated by shuffling all the values of a feature and observing how the model's performance changes. Numerically the importances are the value of the loss function *minus* the value of the loss function after the shuffling.

The related error is estimated statistically through repetitions of the shuffling.

Notice the smart idea behind this procedure. If you shuffle the values of a set of variables, the statistical properties like mean, variance, and so on are retained, but we have destroyed the causal dependence between the targets.

Here the numbers are variations of the loss function. They show by how much the loss function will increase so they provide to us merely a ranking. They don't show to us a relative importance variation: the "Goal Scored" feature is not three times more important than the "Distance Covered" one. It gives three times more variation of the loss function that is not the same thing as to contribute three times more to the answer of the model.

In Machine Learning courses, we have seen how dropping less important features (under a specified value you have to guess) can improve model performance. The reason is usually to drop irrelevant features.

But we must have special attention for the case of two or more correlated features. Think of it, if two features are highly correlated with a thought experiment, think of two features that are copies of each other. The model can indifferently use one feature or the other. So when you shuffle one of the two features, the model will use the other to retain some performance so the importance of the shuffled features will be underestimated. *Permutation Importance will underestimate highly correlated features.*

4.1.2   Permutation Importance on the Train Set

At the bottom of the ranking, we have some negative values. A negative value may sound strange, but it simply states that *the model without those features has an increased accuracy.* Such phenomenon is in fact normal in the training of a model, and excluding those features and then retraining the model anew increase the overall performance.

Wait! A model trained on a dataset shouldn't be using those bad features at all! In fact that's precisely true on the training set not on the validation/test set that poses to the model a novel task. The negative values are a case of bad generalization; so more in detail, the negative values are a form of overfitting.

To double check this hypothesis, we apply Permutation Importance to the train set, and the expectation is that we won't have these negative values. Let's do this quick exercise for confirmation.

```
import eli5
from eli5.sklearn import PermutationImportance
 perm  =  PermutationImportance(my_model,  random_state=1).
fit(train_X,train_y) #A
eli5.show_weights(perm, feature_names = val_X.columns.tolist())

 #A Here we train the Permutation Importance model on the train.
set
```

Out[8]:

Weight	Feature
0.1375 ± 0.0243	Goal Scored
0.0187 ± 0.0156	Attempts
0.0104 ± 0.0132	Free Kicks
0.0104 ± 0.0000	Blocked
0.0083 ± 0.0083	Distance Covered (Kms)
0.0062 ± 0.0102	Pass Accuracy %
0.0062 ± 0.0102	On-Target
0.0042 ± 0.0102	Ball Possession %
0.0021 ± 0.0083	Fouls Committed
0.0021 ± 0.0083	Passes
0.0021 ± 0.0083	Corners
0 ± 0.0000	Yellow Card
0 ± 0.0000	Saves
0 ± 0.0000	Red
0 ± 0.0000	Offsides
0 ± 0.0000	Off-Target
0 ± 0.0000	Yellow & Red
0 ± 0.0000	Goals in PSO

Fig. 4.3 Permutation Importance output on the training set, no anomalous negative values here

It is the same code as before, but we changed the test set with the training set. Here the output (Fig. 4.3):

The table confirms our idea: this time we don't have any negative value on the train set, and "goal scored" is confirmed as the most important feature that affects the output.

This confirms our guess of overfitting as further shown by the changes in the rest of the ranking. In particular, "Attempts" is now at the second place, while it is at the bottom of the ranking performed with the test set (also in red). This is a further evidence of the fact that the ML model is badly using features that are not important to overfit the results. *As general recommendation the feature importance methods should be always applied on the test set.* This exercise has been performed just to deal with negative values and check the overfitting.

4.1.3 Partial Dependence Plot

The main strength of this XAI method is to provide a simple and direct answer about the most important feature. The output is a nice and easy table that can be directly passed to our stakeholders. But it doesn't help no answering the "How": we may be interested or asked to answer how goal scored may change the predictions. Is there any threshold on goal scored to increase the probability of having the Player of the Match prize? Permutation Importance cannot help on this, and in a while we will see how to tackle this further point.

Getting back to the point of answering the "How" instead of "What," we introduce the Partial Dependence Plot (PDP) method to deal with this.

PDP sketches the functional form of the relationship between an input feature and the target; as we will see it can also be extended to more than one input feature. PDP is used on a model that has already been fit, and we will use it to see "how" the predictions are changed by changes in the number of goals (our most important feature as from Permutation Importance). What is performed under the covers by PDP method is to evaluate the effect of changes in a feature over multiple rows to get an average behavior and provide the related functional relationship. It is important to note that averaging may hide a subtlety, the fact that the functional relation may be increasing or decreasing for different rows, and this won't appear in the final result that will show just the "average" behavior. Also in the simplest case interactions between features are not taken into consideration, but we will see the case of a two-dimensional PDP later.

Let's go to the real code to touch with hands what we are talking about:

We consider the PDP for goal scored that, according to our previous analysis, is the most important feature.

```
from pdpbox import pdp, get_dataset, info_plots   #A


feature_to_plot = 'Goal Scored'  #B
  pdp_dist  =  pdp.pdp_isolate(model=my_model, dataset=val_X,
 model_features=feature_names, feature=feature_to_plot)

pdp.pdp_plot(pdp_dist, feature_to_plot)
plt.show()

#A Import from the pdpbox library
#B We select the 'Goal scored' feature
```

With this bunch of lines, we just select the feature we want to analyze (Goal Scored) and pass the info to PDP library to do the job; here are the results (Fig. 4.4):

Nice diagram provided by PDP library, right? On x axis we have the goal scored, while on y axis we have the estimated change in the prediction with respect to the baseline value that is set to 0. The shaded area is an indication of the confidence level.

We see a first interesting outcome: we already got from Permutation Importance that Goal Scored is the most important feature; here we see that we have a strong and positive increase in prediction with 1 goal scored, but after that the trend is almost flat, scoring a lot of goals doesn't change to match the overall prediction of having the "Player of the Match" prize.

Let's do the same exercise but with a different feature, the one that was in the second position in the Permutation Importance ranking, that is, the covered distance: "Distance Covered (km)" (Fig. 4.5).

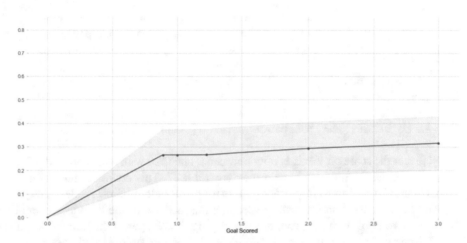

Fig. 4.4 Partial Dependence Plot diagram that shows how "Goal Scored" influences the prediction (Becker 2020)

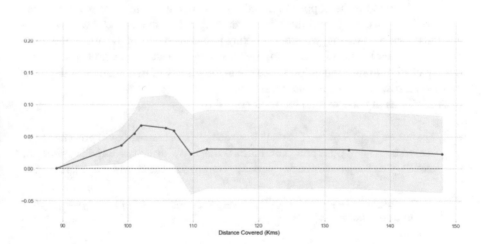

Fig. 4.5 Partial Dependence Plot diagram that shows how "Distance Covered" influences the prediction

```
feature_to_plot = 'Distance Covered (kms)' #A
```

```
 pdp_dist  =  pdp.pdp_isolate(model=my_model,  dataset=val_X,
model_features=feature_names, feature=feature_to_plot)
```

```
pdp.pdp_plot(pdp_dist, feature_to_plot)
plt.show()
```

#A We select the 'Distance Covered (kms)' **feature**

Note that the scale on *y* axis is now between 0 and 0.20 with a max around 0.08, while for "Goal Scored" we have a max around 0.27. This confirms the fact that "Goal Scored" is the most important feature for our Random Forest model.

Also here we have an interesting scenario: the increase of the distance covered has a positive impact on the probability of having the "Player of the Match," but if the distance covered by the team is more than 100 km or so, the trend goes in the opposite direction – running too much decreases the probability for the "Player of the Match" award that was not evident from Permutation Importance analysis only.

Remember that, here, we have just the average effect of one feature on the target; we are not able to see how the values are playing for each predicted results. On average goals are slightly increasing the prediction, but we could have cases (matches) in which the trend is the opposite and others in which the number of goals strongly increased the overall probability; the overall result is the average we see. This depends from the fact that we are not taking into consideration the effect and interplay of the other features but isolating just one. Here we want to exploit another nice feature of PDP library that allows looking at the effects of two features at the same time, to narrow down the mutual interactions.

A few lines of code will do the trick (Fig. 4.6):

```
features_to_plot = ['Goal Scored', 'Distance Covered (Kms)']
inter1  =  pdp.pdp_interact(model=my_model, dataset=val_X,
model_features=feature_names, features=features_to_plot)  #A
```

```
pdp.pdp_interact_plot(pdp_interact_out=inter1, feature_names
=features_to_plot)
plt.show()
```

```
#A  PDP for feature interaction
```

Some new behavior emerges from this diagram compared with the previous two in which we have just one feature per time. The maximum increase of probability for "Player of the Match" prize is with goals between 2 and 3 and a distance covered about 100 km.

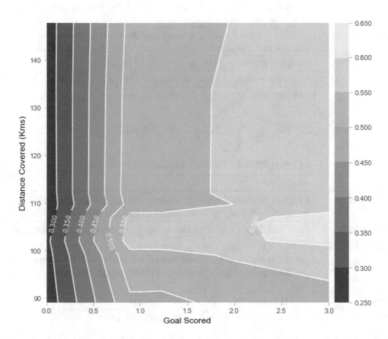

Fig. 4.6 PDP diagram that shows the interaction of the two main features and their impact on the prediction

Looking at the single diagram of goal scored, it seems there is just a slight variation above one goal; here we see that there is a clear area of best values (yellow area). Also, it is confirmed that the maximum effect from distance covered is achieved around 100 km, but with more goals also longer distances produce the same overall effect. We did this exercise just with the two most important features, but you can easily guess that it could be worth to explore also other combinations to deep dive the scenario and produce detailed explanations.

4.1.4   Properties of Explanations

Let's summarize as usual the properties of the explanations we provided (Tables 4.3 and 4.4):

Please compare this table with the analogous one we got for intrinsic explainable models in the previous chapter.

Table 4.3 Explanation property assessment for Permutation Importance and PDP methods

Property	Assessment
Completeness	Interpretability achieved with agnostic method, completeness is low, limited possibility of anticipating model predictions (we can just look at goal scored as rough indicator)
Expressive power	Good in terms of getting evidence of the most important feature but on average and without details of feature interactions (or limited)
Translucency	Low, we don't have insight into model internals
Portability	High, the method doesn't rely on the ML model specs
Algorithmic complexity	Low, no need of complex methods to generate explanations
Comprehensibility	Good level of human understandable explanations

Table 4.4 Properties of explanations for intrinsic explainable models

Property	Assessment
Completeness	Full completeness achieved without the need of trading-off with interpretability being an intrinsic explainable model
Expressive power	Less than linear regression case. Interpretations of coefficients are not so straightforward
Translucency	As any intrinsic explainable model, we can look at the internals. Weights are used to provide explanations but not so directly as in linear regression case
Portability	Method is not portable, specific for logistic regression
Algorithmic complexity	Low but not trivial as in linear regression case
Comprehensibility	Explanations are human understandable also for not technical people

It is evident that for agnostic methods, we are losing completeness (we don't have a full understanding of the model) but gaining portability because agnostic methods are not model-dependent.

It is useful to remark the scope of the explanations we provided; they are global explanations. As we saw we leveraged Permutation Importance and PDP that relies on averages to show functional relationships and explanations. We are not in the position, with these methods, of answering specific questions on a prediction for specific data points. The objective of the next section is to use SHAP to switch to local scope and answer questions on a particular data point prediction.

4.2   Local Explanations: XAI with Shapley Additive exPlanations

Do you remember the beginning of our journey through XAI started in Chap. 1?

The classical example that is always presented to introduce XAI is that someone, say Helen, goes to the bank to ask for a loan but it is refused. The obvious question

Table 4.5 Specific match that will be analyzed with SHAP

Date	Team	Opponent	Goal scored	Ball possession %	Attempts	On-target	Off-target	Man of the match
25-06-2018	Uruguay	Russia	3	56	17	7	6	Yes

that follows is "Why?," and the bank might be in trouble if an opaque ML model has generated the answer without XAI. As global methods, permutation importance and Partial Dependence Plots don't help in this case. Helen is not interested in getting "global" explanations about how the ML model works but wants an answer on her specific case.

If we move to our working scenario of "Player of the Match" prize, so far we provided explanations about the most important features and the functional relationship of these features with the prediction, but we are not able to answer the direct question: considering the features in the figure, how much the specific prediction for this match has been driven by the number of goals scored by Uruguay?

This is the same problem for Helen: she doesn't want to know how the bank ML model generally uses the features but wants to know, in her case, why her loan has been refused.

The same for "Player of the Match": this time, we don't want to know the most important features of the model but get explanations on a specific match, say Uruguay-Russia.

We are transitioning from global explanations to local explanations using SHAP library. Here SHAP stands for Shapley Additive exPlanations (Table 4.5).

From Permutation Importance and PDP, we know that "on average" number of goals is the main driver of the prediction, but we also know (remember the 2D PDP plot) that the features may have mutual interactions that can change the situation. Moreover, here we want to know how much the prediction has "likely" been increased, compared to a baseline, by the fact that Uruguay scored exactly three goals. SHAP helps in these specific cases, where we need an answer on a single prediction and we are less interested in an understanding of the "average" behavior of the model (Becker 2020).

4.2.1 Shapley Values: A Game Theoretical Approach

SHAP method relies on Shapley value, named by Lord Shapley in 1951 who introduced this concept to find solutions in cooperative games. To set the stage, game theory is a theoretical framework to deal with situations in which we have competing players and we search for the optimal decisions that depend from the strategy adopted by the other players. The creators of modern game theory were mathematicians John von Neumann and John Nash and economist Oskar Morgenstern.

Cooperative game theory is a specific case of game theory in which the assumption is to have a group of players that make decisions as coalitions building cooperative behavior. Shapley values deal with this specific scenario: we have a coalition of players in a game that, with a specific strategy, achieves a collective payoff, and we want to know the fairest way to split the payoff among these players according to the contribution that each of them provided.

How to estimate this marginal contribution of each player?

Assuming to have three players Bob, Luke, and Anna who join the game one after the other, the most straightforward answer would be to consider the payoff achieved by each of them:

Bob joined and got a payoff of 7, then Luke joined to bring a payoff of 3, and last Anna added a payoff of 2; so the sequence would be (7,3,2).

But here we are not taking into consideration the fact that changing the sequence in which the players join the game may change the respective payoff (because of the different game background conditions they may find at the time they enter). Also, we need to consider the case in which all the players join the game simultaneously.

Shapley values answer this question by doing an average over all possible sequences to find each player's marginal contribution. We'll see this adding some technical details in Sect. 4.3.

How is all this stuff related to Machine Learning and XAI? The analogy that is adopted is powerful, and it is an excellent example of strong ideas that pass the barriers from one domain of science to another.

We can replace "players" with "features" that are now playing to build the prediction that is our payoff. Shapley values will tell us how the payoff is fairly distributed among the features, that is, which features contributes more for a specific prediction – the outcome of a game.

If you prefer, you can also think of Shapley values as a fair repartition of wages for workers.

Each worker's wage will be proportional to his contribution, and his contribution is calculated precisely via Shapley values.

The XAI method is called SHAP – acronym for Shapley Additive exPlanations – and provides explanations of a single prediction through a linear combination (additive model) of the underlying Shapley values. Let's see how it works in practice.

4.2.2   The First Use of SHAP

Back to our specific question, we have a match "Uruguay vs Russia," and we want to know how much the prediction for "Player of The Match" prize that resulted from the ML model has been driven by the number of goals scored by Uruguay. We compare the prediction to a baseline value that is defined as the average value for all the predictions of all the matches.

```
row_to_show = 19
data_for_prediction = val_X.iloc[row_to_show]   #A
  data_for_prediction_array   =   data_for_prediction.values.
  reshape(1, -1)
my_model.predict_proba(data_for_prediction_array)


#A use 1 row of data here. Could use multiple rows if desired
```

These lines of code are just to select the right match (Uruguay – Russia) and see the prediction from our ML Random Forest model of having Uruguay assigned the "Player of The Match" award (52%):

```
array([[0.48, 0.52]])
```

Following the snip of code to import SHAP library and use it:

```
import shap   #A
k_explainer   =   shap.KernelExplainer(my_model.predict_proba,
  train_X) #B

k_shap_values = k_explainer.shap_values(data_for_prediction)


#A package used to calculate Shap values
#B # use KernelSHAP to explain test set predictions
```

With these three lines of code, we have already produced the SHAP values that can be used to explain the specific prediction for the match, but they would be just numbers. The most exciting feature is the built-in graphic library that allows for a beautiful and interpretable plot of the results.

```
  shap.force_plot(k_explainer.expected_value[1],   k_shap_val-
  ues[1], data_for_prediction)
```

Let's look at the output and how to interpret it (Fig. 4.7):

Fig. 4.7 SHAP diagram that shows how the features impact on the match Uruguay-Russia. A force diagram representing how much the features change the final value. For example we see that "Goal Scored = 3" has the most impact for it pushes the final value to the right with the biggest interval (Becker 2020)

There are two kinds of main indicators in the diagram. Features on the left are the ones that increase the predictions, and their relative length is an indication of the importance of the features in determining the prediction.

Features on the right, same logic, are the ones that are expected to decrease the prediction value. For this specific match, we predicted a probability of 0.52 for Uruguay to have "Player of the Match" that is not so high considering that they scored three goals (remember that goal scored was identified as the most important feature).

The value 0.52 has to be compared with the baseline value of 0.50, which is the average of all the outputs and, in this case, represents the maximal uncertainty. We see the explanation of this result: albeit the red driving features, we have the free kicks, attempts, and off-target features that depress the overall probability. The shift from the baseline is the difference in length between the sum of red bars and the blue bars.

It is important to stress again the big difference with what we did before. With Permutation Importance and PDP, we identified the most important features and provided an "average" functional relationship between these features and the prediction. We had no chance of getting into a specific prediction to answer a why question on a specific match. With SHAP we are able to answer a question on a specific occurrence so to address problems in which a person wants to know what happened with his case (loan rejected) and is not interested in having general explanations about the relative importance of the features: we are interested in what happened to us ONLY! The number of goals scored in the Uruguay-Russia match would make us predict a higher value for the prediction, while for this specific match, SHAP told us that there are other factors that limit the prediction to 0.52.

But SHAP can do even more, and it is not limited to the deep dive of a single prediction.

Do you remember one of the limitations of the Permutation Importance method? We can use it to know the relative importance of a feature, but we don't know if that feature contributed a lot for few predictions and almost nothing for the rest producing an average behavior. SHAP allows getting a summary plot in which we see the impact of each feature on each prediction.

```
shap_values = k_explainer.shap_values(val_X) #A
shap.summary_plot(shap_values[1], val_X)

#A We call the summary plot
```

The code produces the following (Fig. 4.8):

You see on the left the list of features and on x axis the SHAP value. The color of each dot represents if that feature is high or low for that specific row of data. The relative position of the dot on x axis shows if that feature contributed positively or negatively to the prediction. In this way, you may quickly assess if, for each prediction, the feature is almost flat or impacting a lot some rows and almost nothing to the others.

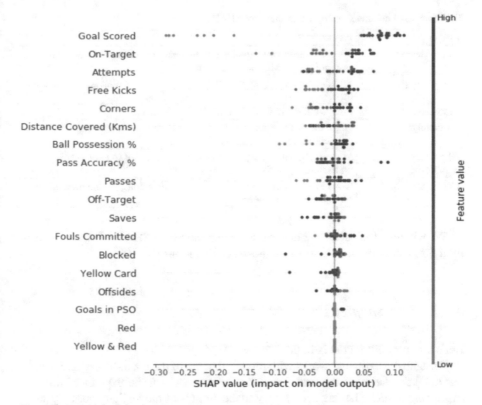

Fig. 4.8 SHAP diagram that shows the features' ranking and the related impact on the match prediction (Becker 2020)

4.2.3   *Properties of Explanations*

Before closing this section, let's do the usual assessment of the explanations we provided.

Let's summarize, as usual, the properties of the explanations we provided (Table 4.6):

After this practical approach, we'll now deepen our understanding of the mathematical foundations of the Shapley values and meet SHAP library and the various algorithms it provides to us.

4.3   The Road to KernelSHAP

To deepen understanding, we briefly discuss the theoretical properties of Shapley values and why SHAP library is the de facto state of the art of post hoc explanations.

Table 4.6 Explanation property assessment for SHAP

Property	Assessment
Completeness	We are focusing on explaining individual predictions; we don't have insight into the ML model machinery
Expressive power	Strong expressive power for the single predictions, that is, what is asked usually to XAI
Translucency	Low, we don't have insight into model internals
Portability	High, the method doesn't rely on the ML model specs to explain the predictions
Algorithmic complexity	Medium, SHAP is easy to implement, but it is not so easy to get a full understanding of the underlying concepts
Comprehensibility	Good level of human-understandable explanations, the diagram that is produced is very powerful

We will also look at LIME method in comparison with SHAP. LIME is an acronym for local interpretable model-agnostic explanations.

4.3.1   The Shapley Formula

Let's say our built model is in the form $y = f(x)$.

An agnostic explainer works with models that are of the black-box type. So without any knowledge of the inner workings of the *model f*, the explainer builds an *explaining model g* having access only to the outputs of model *f* and possibly some info on the training set or the domain.

As we have already said, Shapley values have a deep foundation in game theory – a theoretical base that many other explanation methods lack. In the original foundational papers and subsequent work from Scott Lundberg and others (Lundberg and Lee 2016, 2017), Shapley explanations have two important properties:

(s.1) They are *additive*, so we can share the quantity to explain between the features

$$g(x) = \phi_0 + \sum_{i=1}^{M} \phi_i = \text{a constant contribution} + \text{sum of each feature's importance}$$

where ϕ_i is the contribution of each feature of the model and ϕ_0 is independent of features.

(s.2) They are *consistent*, or we can say *monotonic*, in the sense that if a feature x_i has more influence on the model than another feature x_j, then $\phi_i > \phi_j$.

It has been demonstrated that Shapley values are unique, in the sense that they are the only possible explanations to have properties (s.1) and (s.2). This gives an enormous appeal to such a method.

Now, how can they be implemented? The formula for Shapley values is

$$\phi_i = \sum_{S \subseteq N\{i\}} \frac{(M - |S| - 1)!\,|S|!}{M!} \Big[f_x \big(S \cup \{i\} \big) - f_x \big(S \big) \Big] \tag{4.1}$$

Here we sum on every possible subset S of features not including the feature i we are investigating. So $f_x(S)$ is the expected output given the feature subset S, and $f_x(S \cup i) - f_x(S)$ is thus the contribution made by adding the feature i.

The combinatorial is a weighting factor that takes into account the multiple ways of creating subsets of features.

4.3.2   How to Calculate Shapley Values

Wait! How can we evaluate f without using some features? We have to use some background information to evaluate f with fictitious values instead of the features we are investigating. It is not a trivial matter, and usually, we use some data distribution, or we give to the method a background dataset to sample from randomly.

The number of possible subsets of features not including feature i is 2^{N-1}, where N is the total number of features, so it increases very quickly with the number of features.

A standard calculation for large features is practically too time-consuming, so we can think only of an estimate approximating it via a Monte Carlo (random) approach.

The work of Lungren, implemented in the SHAP library, shows to us other attractive solutions (and speedups).

KernelShap used in the previous section is an agnostic *approximate* linear approximation and works for every possible model you may train.

TreeShap is not agnostic because it only works on tree-based models (even boosted trees), but it works in *linear* time, and it is even an *exact* calculation of Shapley values not an approximation.

There are also some specific methods for *Deep Neural Networks*, but we will see them in the following chapter.

4.3.3   Local Linear Surrogate Models (LIME)

The idea behind KernelShap is to construct a local surrogate model for the explanation model g. A surrogate model is an effective approximation of the model: a reconstruction of the model that can give approximately the same results of the model. A local surrogate model will approximate the real model for values near to a given sample.

As we have already seen, a local explanation model is more powerful than a global one. It can answer the customer's question "Why my loan has been refused?"

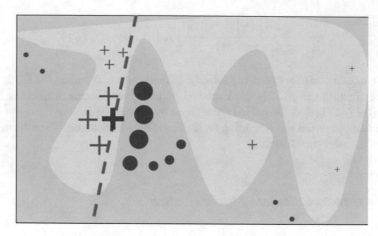

Fig. 4.9 Schematics for LIME. The class outputs of the model are circles or crosses, and the dimension reminds us of the weight, so distant points are weighted less (Ribeiro et al. 2016)

probing all the answers that the model will give for little variations of customer data. Technically it gives different explanations for each instance of the model.

The requirement of additivity (s.1) in the previous subsection forces us to use a locally linear model, which, in fact, is an excellent idea for we know linear models are intrinsically explainable. So it turns out that KernelShap is not so different from another well-known agnostic technique called LIME.

The idea of the original LIME by Ribeiro et al. (2016) (as we said LIME stands for local interpretable model-agnostic explanations) is to find a surrogate linear model repeatedly calling the trained model f. Say f is a classifier: the value $f(x)$ is the probability of a class for the instance x (Fig. 4.9).

To construct a linear model g, we add some Gaussian noise to x to have some new perturbed points say z_1, z_2, z_3, and so on. We call the model f on these new instances to have new class probabilities $f(z_1)$, $f(z_2)$, $f(z_3)$, and so on. Finally we train a linear g on these new instances with the requirement that distant points are weighted less using a weight exponentially decreasing with the distance.

This process can be described with a loss function of the parameters of g with two terms:

$$\text{Loss}(g) = L(f, \pi, g) + \Omega(g)$$

$L(f, \pi, g)$ is the usual sum of squared differences between the values' instances $f(z_i)$ and the surrogate model g you would expect from a loss function. But here it's multiplied with a positive weight factor π decreasing with the distance from the original instance x.

In LIME, π is usually a decreasing exponential function.

The new term $\Omega(g)$ is a Lasso regularization to have a sparse representation. We can change $\Omega(g)$ to reduce the dimension of the explanation to say only K non zero features.

Fig. 4.10 Why do you say this is a wolf? Because we have snow! (Ribeiro et al. 2016)

When we restrain only to a few nonzero features (without losing the fidelity of our explanation), we increase simplicity and so we gain in interpretability.

In fact, in literature, usually, we differentiate the instance x (which like all the features of the model f) with its interpretable analog x' where only the meaningful characteristics are considered.

Think of the picture of the wolf we have already seen in the first chapter. In this case, x is a matrix with all the pixel values (in the three colors) while x' is only the selected region with snow (Fig. 4.10).

4.3.4 KernelSHAP Is a Unique Form of LIME

Now that we know how a linear surrogate is constructed, we can state the following striking property: KernelShap is the only local linear surrogate explanation model g of the initial model f that gives us Shapley values.

In KernelShap, we use the LIME loss but weighting by a "distance" π counting the number of possible subsets of features we use in Shapley formula (4.1).

So, as a matter of fact, KernelShap is a special type of LIME.

We conclude by showing remarking one advantage of SHAP.

Shapley values are all of the same dimensions (say dollars) even if the corresponding features are not. So we could profitably use Shapley values themselves as new features for another model without using normalization. Or we can group different samples together using the similarity of their Shapley explanation. Technically, Shapley values are beautiful feature extractors.

SHAP library includes even two models adapted to explain Deep Neural Networks, but we'll see them in the next chapter.

4.4   KernelSHAP and Interactions

4.4.1   The New York Cab Scenario

Let's start with our scenario that deals with Kaggle dataset for the New York Cab dataset (Kaggle 2020). The objective is to predict the fare amount for cabs in New York based on pickup place and drop-off locations. The estimation is very basic and takes into consideration only the distance between the two positions.

We will go again through SHAP but with more care toward interactions of features; also, we will assume to have a critical request on *providing timely explanations*, and to achieve this performance goal we will record prediction times.

4.4.2   Train the Model with Preliminary Analysis

Let's open the file

```
import pandas as pd
import numpy as np
import matplotlib.pyplot as plt
from lightgbm import LGBMRegressor   #A
from sklearn.model_selection import train_test_split
from sklearn.metrics import r2_score

from sklearn.inspection import permutation_importance #A


# Data preprocessing.
data =  pd.read_csv("./smalltrain.csv", nrows=50000)

#A We will use Sklearn
```

Now we filter outliers and train a gradient boosting model

```
data = data.query('pickup_latitude > 40.7 and pickup_latitude <
40.8 and ' +
      'dropoff_latitude > 40.7 and dropoff_latitude < 40.8 and ' +
      'pickup_longitude > -74 and pickup_longitude < -73.9 and ' +
      'dropoff_longitude > -74 and dropoff_longitude < -73.9 and '
+
      'fare_amount > 0'
      )
```

```
y = data.fare_amount


base_features = ['pickup_longitude',
                 'pickup_latitude',
                 'dropoff_longitude',
                 'dropoff_latitude',
                 'passenger_count']

X = data[base_features]


X_train, X_test, y_train, y_test = train_test_split(X,y,test_
size=0.5, random_state=1111)


# Tain with LGBM Regressor
reg = LGBMRegressor( importance_type='split',   random_state=42,
num_leaves=120)   #B

reg.fit(X_train, y_train)
print(r2_score(y_train,reg.predict(X_train)))
print(r2_score(y_test,reg.predict(X_test)))

#A We will train a LGBM Regressor. Note that LGBM is a tree-based
model.
#B Here we train the LGBM Regressor
```

We get an accuracy of

```
0.6872889587581943        train
0.4777749929850299        test
```

Not a good model because the R2 score is pretty low, and we can say it's even overfitting because the score on the train set is significantly larger than that on the test set.

We can calculate the usual Permutation Importance (Fig. 4.11):

```
# Getting permutation importance.
result = permutation_importance(reg, X_test, y_test, n_repeats=10,
random_state=42)
perm_sorted_idx = result.importances_mean.argsort()


# Visualize two variable importance plots.
fig, ax1= plt.subplots(1, 1, figsize=(12, 5))
```

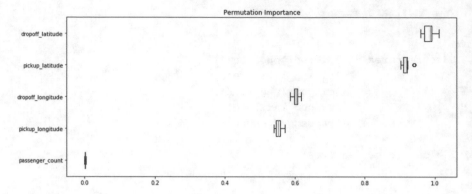

Fig. 4.11 We plot the Permutation Importance for the New York Cab dataset. We use a boxplot to express the uncertainty of the Importance estimated repeating n_repeats times the calculation

```
ax1.title.set_text('Permutation Importance')
ax1.boxplot(result.importances[perm_sorted_idx].T, vert=False,
            labels=X_test.columns[perm_sorted_idx])

fig.tight_layout()
plt.show()
```

As we have already seen, Permutation Importance shows to us the features that have a major impact on the loss function as in the *y* axis of the figure.

Now we make an "interaction PDP" plot between the two more important features (following the Permutation Importance) (Fig. 4.12):

```
print('Computing partial dependence plots...')
from sklearn.inspection import plot_partial_dependence
import time

tic = time.time()
fig, ax = plt.subplots(figsize=(5, 5))
plot_partial_dependence(reg, X_test, [(X_test.columns[0],X_
test.columns[3])],
                        n_jobs=3, grid_resolution=20,ax=ax)
print("done in {:.3f}s".format(time.time() - tic))
ax.set_title('Partial dependence of NY taxi fare data - 2D')
plt.show()

With output


Computing partial dependence plots...
done in 21.406s
```

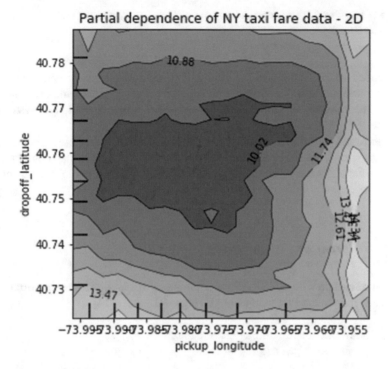

Fig. 4.12 Partial Dependence Plot for the two most important features

Again the picture can show us how the two features are playing together.

When two features do not interact, we can write $f(x) = f_j(x_j) + f_i(x_i)$, so each feature contributes independently from the other. And the picture suggests to us a complex interacting behavior like in the Player of the Match scenario.

4.4.3   Making the Model Explainable with KernelShap

Now we search for local explanations using SHAP library

```
import shap
tic = time.time()

background=shap.kmeans(X_train, 10)    #A


explainer = shap.KernelExplainer(reg.predict, background)    #B
shap_values = explainer.shap_values(X_test, nsamples=20)
print("done in {:.3f}s".format(time.time() - tic))
```

```
#A We need background information. For speed we'll summarize
the background as K = 10 samples
#B Using KernelSHAP
```

To speed up the creation of the model, we used a little trick instead of passing as a background, a big set of samples; we used k-means to reduce the example set to only 10 meaningful centroids. Nonetheless, model creation took *115.090s* on our machine.

As introduced at the beginning of the section, in this scenario, we are assuming that explanations should be available as fast as possible. KernelShap computation is too slow for tasks where speed is a must. We will see in a while how to improve performance.

4.4.4   Interactions of Features

Now for reference with our SHAP model trained, we can do feature importance and a Partial Dependence Plot using SHAP (Figs. 4.13 and 4.14)

```
# Variable importance-like plot.
shap.summary_plot(shap_values, X_test, plot_type="bar")

shap.dependence_plot("pickup_latitude", shap_values, X_test)  #A
```

#A Let's make a PDP-like plot with SHAP

Here every point is a different sample. On the left vertical axis, we have the SHAP value for that sample for the feature pickup_latitude and on the horizontal axis the pickup_latitude_value. The model automatically finds the feature that is

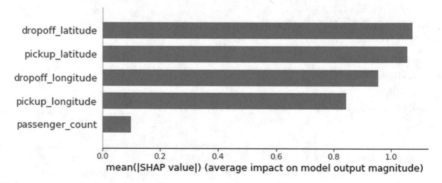

Fig. 4.13 Feature Importance diagram using SHAP

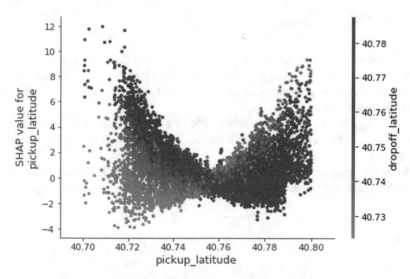

Fig. 4.14 Partial Dependence Plot using SHAP

most likely interacting with "pickup_latitude" and uses these features to color the points. On the right vertical axis, we read that the interacting feature is dropoff_latitude.

If the two features were not interacting, the overall coloring would be uniformly distributed, or the shades of a different color would be not intersecting.

We see instead a definite pattern with an intersection showing us a relevant interaction. What the picture roughly shows is that when dropoff_latitude is low (Lower Side of NYC), the contribution of pickup_latitude to prediction (i.e., the SHAP value of pickup_latitude) is increasing the fee. Instead when dropoff_latitude is high, it seems that the pickup_latitude contribution to the fee is decreasing with the increase of the pickup_latitude variable. As a matter of fact, the feature dropoff_latitude is changing how the pickup_latitude feature contributes to the fee.

4.5   A Faster SHAP for Boosted Trees

As mentioned, TreeShap contains a faster version for tree-based models as well as for the Lightgbm model we used in the previous section. Also, TreeShap is not only faster but also exact in calculation of Shapley values.

4.5.1   Using TreeShap

The idea here is calculating the Shapley values as a weighted average at every node of the Shapley contribution of the branch. The algorithm, which does a clever reuse of previous values to collect results, can even estimate the background contribution from the tree structure.

Obviously, TreeShap is no more agnostic like KernelShap for it uses the model inner structure, but we have a wonderful trade-off both in speed and precision of calculation.

So let's retrain, on the same boosted model before, with TreeShap instead of KernelShap to meet the requirement of shortening the time to provide explanations

```
import shap
print('Computing SHAP...')
tic = time.time()

explainer = shap.TreeExplainer(reg) #A


shap_values = explainer.shap_values(X_test)
print("done in {:.3f}s".format(time.time() - tic))
        pd.DataFrame(shap_values,columns=X_test.columns)
```

```
#A  Using TreeSHAP, not KernelSHAP. Remember reg is a LGBM model
that is a tree-based model.
```

with output:

```
Computing SHAP...
 Setting feature_perturbation = "tree_path_dependent" because
 no background data was given.
done in 18.044s
```

It took only *18 seconds* compared with the previous result of *115 seconds*, and this is a huge improvement! Our stakeholders who were pushing to get timely explanations about the predicted fares will definitely be happy.

4.5.2   Providing Explanations

To complete the work with The New York Cab Company (or any agency interested in the control/taxation of fares), we have to show some examples and the SHAP at work with a force plot (Fig. 4.15).

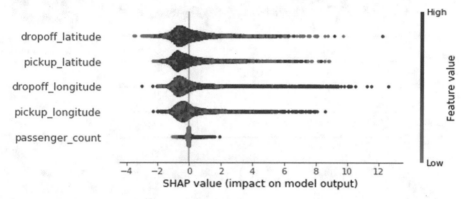

Fig. 4.15 SHAP diagram for cab scenario

Fig. 4.16 SHAP summary plot for cab scenario

```
shap.initjs() # print the JS visualization code to the
notebook
# visualize the a prediction's explanation, decomposition
between average vs. row specific prediction.
shap.force_plot(explainer.expected_value, shap_values[50,:],
X_test.iloc[50,:])
```

We readily see the positive and negative contributions at the fare. Now, what if we want to see all the force plots for all the samples at once? We can have a summary of where each point is a different sample (Fig. 4.16).

```
# Each plot represents one data row, with SHAP value for each variable,
# along with red-blue as the magnitude of the original data.
shap.summary_plot(shap_values, X_test)
```

Or we can rotate vertically the force plots and pack them horizontally to have one plot

```
# Pretty visualization of the SHAP values per data row. We limit
to the first 5000 samples
          shap.force_plot(explainer.expected_value,   shap_val-
ues[0:5000,:], X_test)
```

sorting them by pickup_latitude (Fig. 4.17).

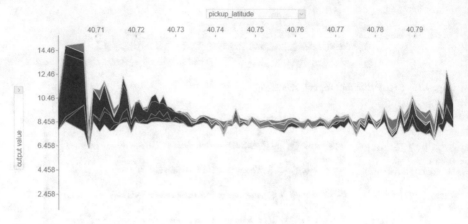

Fig. 4.17 Packed SHAP force plots

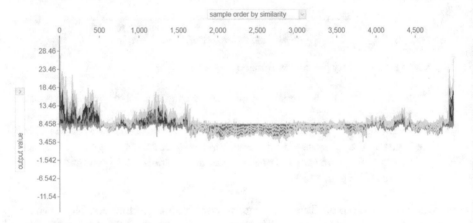

Fig. 4.18 SHAP diagram: similarity between explanations

Or grouping SHAP explanation using the similarity between the explanations (Fig. 4.18).

4.6   A Naïve Criticism to SHAP

We conclude with a clever objection by Edden Gerber (2020) who proposes an explanation method inspired by SHAP named Naïve Shapley values.

The method of Gerber, albeit with its limits, clarifies what we truly find in SHAP in contrast to what we would expect to find in SHAP.

If we return to formula (4.1), we see that crucial to the calculation of Shapley values is f_x, the evaluation of the model *without* some specified features.

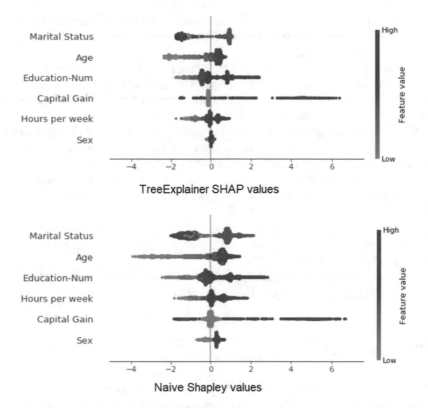

Fig. 4.19 The adult census database – UCI Machine Learning Repository (UCI 1996) – SHAP documentation

As we know, SHAP method replaces the missing features with some background information (or statistics) about the missing features, but what if the model f_X would be already independent of the missing features?

In fact, Gerber's Naïve Shapley values retrain a model f_X for each coalition of missing features and after that calculate Eq. (4.1).

This methodology is not new, it is just a form of bagging a technique already at the core of Random Forest, and it is not an agnostic method anymore for it does not explain the original model f_X but retrains a bunch of new reference models.

But it can really be of use in the modeling phase, and it's results will surprise us.

Now if we have to retrain the models, we must have access to the same train set used to train the original model. We possibly augment the model with predictive power it originally has not, and the training time grows exponentially with the number of features involved.

In fact Naïve Shapley describes the properties of the data more than the model f_X we want to explain but the results are somewhat enlightening.

We start from training a model f on the adult census, a database already included in the SHAP library predicting the annual income of people (Fig. 4.19).

In the picture, we have both the explanation of TreeExplainer SHAP on the model *f* and the description of Naïve Shapley values of the same model plus the accessory trained models.

We see that the values are very similar, but the Naïve have more spread out values.

Look at the *Gender* feature; for instance, the original model can't clearly show the effect of Gender on the income, but Naïve Shapley values can because there is gender disparity in the data.

We have already seen this effect in the section on Permutation Importance; the original model has learned to use other features than gender to predict income so it thinks gender is less important in the prediction than it is in the data.

Instead "Naïve Shapley" retrains other models that are forced to use gender when some other features are missing, so it is more representative of the real data, and it doesn't explain well what the model does.

As a matter of fact, we could train better models in terms of fairness comparing the trained model's SHAP values with the data analysis provided by the "Naïve Shapley values" approach.

In the next chapter we will introduce the explainable models for Deep Learning. And we will literally try to open the (black) box!

4.7   Summary

This chapter has been devoted to model-agnostic methods starting with Permutation Importance and Partial Dependence Plots before deep diving into the more complex Shapley values and SHAP methods:

- Use Permutation Importance to answer "What" questions on the most important features.
- Use Partial Dependence Plots to answer "How" questions to understand the impact of the features on the predictions.
- Provide local explanations using SHAP.
- Enrich our understanding of SHAP to be compared with LIME.
- Improve performance on generating explanations using TreeShap instead of KernelShap.
- Get awareness on the limits of SHAP to tailor and adapt the best XAI strategy for a specific case.

References

Becker, D. (2020). *Machine learning explainability*. Available at https://www.kaggle.com/learn/machine-learning-explainability.
Gerber, E. (2020). *A new perspective on Shapley values, part II: The Naïve Shapley method*. Available at https://edden-gerber.github.io/shapley-part-2/.

Kaggle. (2020). *New York taxi fare prediction*. Available at https://www.kaggle.com/c/new-york-city-taxi-fare-prediction.

Lundberg, S., & Lee, S. I. (2016). An unexpected unity among methods for interpreting model predictions. In *29th conference on Neural Information Processing Systems (NIPS 2016)*, Barcelona, Spain, pp. 1–6. http://arxiv.org/abs/1611.07478.

Lundberg, S. M., & Lee, S. I. (2017). A unified approach to interpreting model predictions. In *Advances in neural information processing systems* (pp. 4765–4774). US:MIT Press

Ribeiro, M. T., Singh, S., & Guestrin, C. (2016). "Why should I trust you?" Explaining the predictions of any classifier. In *Proceedings of the 22nd ACM SIGKDD international conference on knowledge discovery and data mining* (pp. 1135–1144).

UCI. (1996). *Adult data set*. Available at https://archive.ics.uci.edu/ml/datasets/adult.

Chapter 5
Explaining Deep Learning Models

"The sculpture is already complete within the marble block, before I start my work. It is already there, I just have to chisel away the superfluous material."
—Michelangelo

This chapter covers:

- Occlusion.
- Gradient models.
- Activation-based models.
- Unsupervised activation models.
- Future prospectives.

  – Provide explanations for Deep Learning models in computer vision.
  – Build an explainable agnostic model for a black box in computer vision.
  – Use saliency maps to provide explanations focusing on regions of major interest.
  – The reader will have a glimpse of the future of the field.

      Build interpretable CNN.
      Use unsupervised learning to do exploratory analysis on a model.

In this chapter, we will talk about XAI methods for Deep Learning models.

The explanation of Deep Learning models is a matter of active research, so we will illustrate the criticalities and advantages of what are the methods of today and could be those of the future.

In this chapter, we want to stress a fundamental concept: having an explainable model is a way to create a robust and reliable model. So there is no need for trading between explainability and robustness; on the contrary, explainability naturally makes the model more robust.

© The Author(s), under exclusive license to Springer Nature Switzerland AG 2021
L. Gianfagna, A. Di Cecco, *Explainable AI with Python*,
https://doi.org/10.1007/978-3-030-68640-6_5

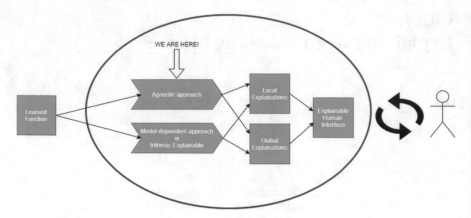

Fig. 5.1 XAI flow: agnostic approach

We will see that XAI methods and training best practices use similar methodologies, and any methodology that speeds up the training of a Deep Learning model and makes it more accurate overlaps with a XAI method.

For simplicity, we will make examples taken from computer vision, and we will face the problem of creating a "good explainable model": we will start from an agnostic point of view, and then gradually, we will try to open the black box leveraging the more advanced techniques in this area (Fig. 5.1).

5.1   Agnostic Approach

We begin our discussion with the agnostic approach, which is the approach in which we perturb our model's input without exploiting the internal functioning of Deep Learning models. Only in the following sections we will cover gradient-based methods.

5.1.1   Adversarial Features

It is not possible to tackle the discussion of Deep Learning models without properly framing the problems and difficulties they entail.

For concreteness, we begin with a series of conceptual experiments.

Let's think about the most classic classics, a model to classify dogs and cats' images.

By treating the model as a black box, we do not care at all if the model is a neural network, a decision tree, or even a linear model. We do not care at all if the model has a preprocessing of the image. If you internally transform it to

Fig. 5.2 Adversarial features in images. (Courtesy of L.Bottou (2019) for the design)

grayscale, do some feature extractions, image convolutions, or so on. Think of the model in the abstract, without any look at its implementation.

Surely you will think that it is enough to have images that have been correctly labeled to solve the classification problem. Yet a question that often goes unnoticed is that of the origin of the picture.

Let's take two images, one of a dog and one of a cat (Fig. 5.2).

We do not go into the details of the camera used, but obviously, we tend to photograph dogs in the open air and cats indoors.

Dogs are usually stationary, and cats are always *on the move*, so we will have to use different exposure times. Also, dogs and cats have different sizes, and therefore, magnification will also be different.

An excellent black box can exploit this information from the photo through the SNR, color balance, depth field, and noise of texture. Thus a good black box can learn to distinguish dogs and cats from features that are absolutely not the features used by a human being.

We have already met the example of the classification of the wolf through the snow. Here we are dealing with the same problem, but we are trying to generalize the concept. Many of the features we have listed are legitimate features naturally occurring in the photos.

In the following, we will call adversarial features, those features that distort learning, typically reducing the ability to generalize the black box.

It is easy to imagine that a classification carried out by the field of view alone, for example, would classify a pony as a dog.

The training of a black box for image recognition must therefore include pre-processing so that the model does not learn from adversarial features. The most used approach to limit this risk is to randomize the image sources as much as possible and resort to augmentations.

5.1.2   Augmentations

Augmentations are a very common approach in computer vision to train models.

It is simply a matter of varying the input images through a series of transformations: rotations, translations, enlargements, color and contrast variations, vertical and horizontal scaling, flipping, and so on. They are so common in computer vision that every Deep Learning framework has its own built-in augmentations.

Clearly presenting every possible rotation of a photo to a model in the training phase will force the model to use features that do not depend on the orientation. So a cat will remain a cat even if viewed upside down.

Traditionally a correspondence has always been made between the amount of data and the complexity of a model.

So, for example, complex models of high capacity in the sense of *Vapnik-Chervonenkis dimension of Vapnik and Chervonenkis (1968)* with many parameters require large amounts of data for their training, and the meaning of augmentation would be to increase this amount of data so as not to make the model overfit.

When we talk about Deep Neural Networks, we will see that this approach, luckily, is not completely correct.

Both for regularizations that are imposed on the weights and for phenomena such as the double descent (Nakkiran et al. 2019), for now, let's not talk about architecture. Let's just say that augmentations are useful for increasing the database available in the training phase, but they must not be done in a completely random way because they could actually confuse the model by adding noise. They should be aimed at providing the model with robust information.

And this allows a more generalizable training and often more and often faster.

Let's clarify these concepts with another mental experiment.

5.1.3   Occlusions as Augmentations

Suppose we want to train a black box to recognize oranges and apples (Fig. 5.3).

The black box could learn to distinguish the two fruits from the shape, from the texture, or, for example, from the stem.

The fruit petiole is an absolutely legitimate feature that uniquely identifies the fruit, but it is not a robust feature in the sense that it does not give us robust training. In fact, it is enough to turn one of the two fruits to make the petiole disappear and prevent the model from making its prediction.

It should be noted here that the robustness of a feature is a different concept from the feature importance that we talked about in previous chapters.

Feature importance is the confidence that a model, that has already been trained, has in the importance of a feature and how much the feature is relevant for the calculation of the output.

Fig. 5.3 How do we distinguish an apple and an orange from the skin's texture, color, or stem? (Photo by vwalatke)

Now let's take two models: one that has learned to distinguish fruit from petiole and one that has learned to distinguish fruit based on peel; both consider respectively petiole or peel as very important features, and both models have the ability to correctly predict the fruit, but the model that uses the petiole is less robust, so in front of new examples it will be less generalizable.

We talk about overfit when a model fails to generalize well on new cases, so the robustness of a model is to build a model that, in some sense, has not overfitted in learning from its training set.

There are several techniques we can use to force a black box to learn to distinguish oranges from the skin and not from the stem.

One could be to provide the model with images with different levels of blur in the image in order to make it less easy to use the stem as a feature, but this method is definitely destructive for the image.

Another technique that can be used is that of *occlusion* (Zeiler 2013) in which some information has been cut off from the images: as an example, a small gray square patch could be randomly applied to the image (Fig. 5.4).

The stem has been cut off to force the model to walk the path of recognition through the peel. This does not reduce the accuracy of the model but increases its robustness and generality.

5.1.4   Occlusions as an Agnostic XAI Method

This is a fantastic example of correspondence between good training and XAI because, starting from the occlusion, a XAI method can be built to understand what the black box is looking at.

Fig. 5.4 The occlusion idea as an augmentation technique: random gray rectangles force the model to rely more on robust features such as skin's texture

If using the occlusion in the training phase, we forced the black box not to learn by looking at the finer details; now, we can take a pre-trained black box and question it on image content using occlusions as a XAI method.

The model has already been trained and fixed, so we don't care in this phase of training it in a robust way; we want to understand which details of the image are most significant for the class' attribution or to evaluate the importance of some groups of pixels.

With the application of the occlusion, we have varied the input, so a priori we will have a different output.

An analysis with occlusions is equivalent to sliding the patch square in each position along the image and assessing how much the output is distorted from time to time.

This is the most classic implementation of the algorithm, which obviously will take longer the smaller the patch. A slightly smarter solution is to approach the problem as an optimization problem and apply a metaheuristic algorithm to quickly evaluate the areas where the square patch will give larger differences for the expected class.

The result of the analysis will be a heatmap or a saliency map that will indicate the most sensitive parts of the model.

And this analysis is independent of the type of model or its internal functioning, so the explanation is agnostic.

But now, let's try to apply it to a real model. We take a pre-trained model in Keras and visualize the most important parts using a library.

In `Keras`, the best-known libraries are `tf-explain`, `DeepExplain`, and `eli5`, while for `PyTorch` users, there is the beautiful `Captum`.[1]

---

[1] https://github.com/sicara/tf-explain
https://github.com/marcoancona/DeepExplain
https://eli5.readthedocs.io/en/latest/tutorials/keras-image-classifiers.html
https://github.com/pytorch/captum

Here we will use `tf-explain` before we install the library by using pip

```
pip install tf-explain
```

then we load the libraries

```
import tensorflow as tf
from tf_explain.core.occlusion_sensitivity import
OcclusionSensitivity
```

and we import a pre-trained ResNet50 model on the ImageNet dataset

```
if __name__ == "__main__":
    model = tf.keras.applications.resnet50.ResNet50(
        weights="imagenet", include_top=True
    )
```

For the rest of the code, we load a specific image and launch two instances of the explainer, one to identify the areas of the image that most likely provide the class tabby cat and the other to identify the areas that most indicate the dog class (Fig. 5.5):

```
IMAGE_PATH = "./dog-and-cat-cover.jpg"
    img = tf.keras.preprocessing.image.load_img(IMAGE_PATH,
    target_size=(224, 224))
    img = tf.keras.preprocessing.image.img_to_array(img)

    model.summary()
    data = ([img], None)

    tabby_cat_class_index = 281
    dog = 189

    explainer = OcclusionSensitivity()
    # Compute Occlusion Sensitivity for patch_size 10
    grid = explainer.explain(data, model, tabby_cat_class_index, 10)
    explainer.save(grid, ".", "occlusion_sensitivity_10_cat.png")

    # Compute Occlusion Sensitivity for patch_size 10
    grid = explainer.explain(data, model, dog, 10)
    explainer.save(grid, ".", "occlusion_sensitivity_10_dog.png")
```

This is the original image and following the explanations for the tabby cat and common dog classes, respectively (Fig. 5.6).

The parts in yellow are the most important.

Fig. 5.5 Original image

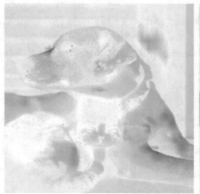

Fig. 5.6 Occlusions used to highlight relevant features respectively for the tabby cat and the dog classes

Since the model of explanation is the occlusion, we can interpret these images by saying that the model needs to look at the textures of the eyes and head to recognize if there is a dog or a cat in the photo. So we can say that, relative to other pixels in the image, those are the most important pixel for the explanation.

We do not doubt that the techniques for the agnostic explanation will become more and more refined and faster in the near future, for example, allowing us to better weigh the joint presence of different classes in the same image.

In the next paragraphs, we will go deeper in terms of explaining and speeding up the process by abandoning the black-box paradigm and opening the model. This will allow us to use differential methods and see the information stored in the model.

5.2   Neural Networks

In this section, we will make a brief review of what neural networks are and how they work. The inner working will introduce us to differential methods.

5.2.1   The Neural Network Structure

A neural network is a Machine Learning model that mimics the functioning of the brain in a simplified model.

Mathematically, this idealization takes place through a graph in which information starts from the input features and arrives at the output features by crossing computational nodes. The most used NN scheme is feedforward or DAG (direct acyclic graph), where information flows from input to output without any loop (Fig. 5.7).

In this respect, a (non-recurring) neural network is undoubtedly different from a brain because in the brain, the connections are also recurrent, and there are different types of memory, short-term and long-term.

Each computational node is typically a linear combination of the inputs then passed through a nonlinear activation function, typically a ReLU function.

This nonlinearity guarantees that the model as a whole is not simply a linear combination of the input features, but, as we will see also in the seventh chapter, the output will remain locally mainly linear in respect of the weights.

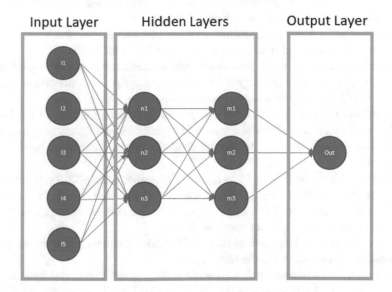

Fig. 5.7 A generic feed-forward network/DAG

Typically the nodes (neurons) are arranged in successive layers. A network with only two layers, the internal (hidden) one and the output one, is called a shallow network. A neural network with two or more internal layers is called Deep.

The current substantial development of Deep Learning is based precisely on the enormous capabilities of deep networks to learn and generalize patterns.

What makes a neural network so efficient?

Layer topology allows us to imagine a neural network as a sequence of several computational blocks. Let's take a shallow network; the transition from the internal state to the output occurs simply through a linear combination. So we can imagine the shallow network as a linear model that does not have, as input, the original features but a convenient transformation of the features into a suitable space of basis functions.

If it is possible to train such a model and if the basis functions are flexible enough, the internal representation (latent representation) assumes a feature extractor's role.

That is, it can transform the initial dataset (an arbitrarily complex task) to a simpler one. So the recognition of an image behaves in the last layer like a standard logistic regression: the highly nonlinear problem is now mapped to a linear one.

It is remarkable that a neural network can extract its features by itself and training a neural network, for example, through backpropagation, is equivalent to simultaneously training the classifier and the feature extractor.

The Universality Theorem of Cybenko (1989) and Hornik (1991) assures us that shallow neural networks are universal approximators of any continuous function. The problem (NP-hard) is to find the exact values of weights to obtain such an approximation.

In practice, for a large class of NN, the search for weights (optimization) through optimizers such as SGD (stochastic gradient descent) and Adam (adaptive moment estimation) is often surprisingly fast and effective.

The breakthrough for the transition to computer vision was discovering a layer type particularly suited to image processing: *convolutional filters*.

Convolutional filters make learning robust to translations of the objects in the images and reduce noise in images. Also, there are far fewer connections between layers in respect of fully connected layers, so fewer weights to train.

Furthermore, you can think of convolutional filters as an analog of the Fourier transforms: they immediately allow you to find periodicity and patterns in the image.

The layered structure of the neural network allows for successive levels of abstraction.

We can think of the first layer as a sort of classifier that splits the points in the image, so the first layer "sees" straight shapes and uniform colors, the next layer "sees" angles, and so on. Each layer responds to (i.e., classifies) more subtle features in the images giving more complex forms.

For this reason, we find neurons that tend to differentiate their functionality more and more the closer we get to the final layers.

Training a neural network using gradient descent is equivalent to propagating the recognition error that is shown in the last layers backward in the network through partial derivatives and chain rule.

Let's take a network of only two layers: the input, the inner layer (also called *hidden layer*), and the layer of the output.

The input will be represented by a vector x. A linear combination will transform the input into the first layer of neurons. A nonlinear sigma activation function (or a ReLU) will be applied to model the response of each neuron.

The output of each neuron will be mapped to the second layer (the output), and a new activation function will be applied (typically softmax) in a classification problem.

In the formula, all two layers of neural networks can be simply expressed as

$$y = \sigma\left(W_2\, \sigma\left(W_1 x\right)\right) \tag{5.1}$$

for suitable weight matrices W_1 and W_2.

In analogy with the logistic regression, the choice activation function σ was historically a sigmoid. In more recent times, the sigmoid has been replaced by the ReLU for speed and convergence of the backpropagation method.

For fun, the reader can refer to https://playground.tensorflow.org.

5.2.2 Why the Neural Network Is Deep? (Versus Shallow)

At this point, it is essential to return to the distinction between shallow and deep networks. Historically the difficulty of training deep networks with only the SGD coupled with the Universality Theorem of Cybenko and Hornik that states that even shallow networks are universal had discouraged the widespread of deep networks.

Furthermore, from the point of view of explainability, a shallow network is almost linear so it can be easily explained with the methodologies we have already dealt with.

So, why move to deeper networks?

Let's take the example of the parity function.

The parity function is a function that has N bits in input and one output. The output is 0 if we have an even number of 1 in input, or it is 1 if we have an odd number of 1.

It can be shown that using a neural network with logic gates, a shallow network can exactly reproduce the parity function with 2^n neurons in its intermediate layer. Instead, a deep network of n layers can reproduce the parity function with just one neuron per layer and, therefore, n neurons.

So Deep Neural Networks are more compact, and this compactness is essentially due to the fact that each layer learns the next level of abstraction.

For this reason, Deep Learning people usually say that shallow networks don't learn the function but merely approximate it.

This property is matched by a fundamental property of deep networks. Computational complexity in training a deep network decreases exponentially with the depth of the network. We call this property Bengio's conjecture. You can read a proof in Mhaskar et al. (2019).

Of course, we are talking about ideal networks, but, from a theoretical point of view, having the possibility to train enormously deep networks allows us to circumvent or mitigate the curse of dimensionality in model training.

The summary is that we cannot hope that the networks remain shallow for the sake of interpretability; instead, Deep Neural Networks are here to stay and become even bigger and deeper (like in GPT-3), so we need to find new ways to explain them.

The good news is that we expect that in the future, the explanation of Deep Neural Networks will cope with the successive levels of abstraction.

For fun, you can visualize the ResNet50 NN we have explained in the previous paragraph in its 25.5M parameters' glory.

```
from tensorflow.keras.utils import plot_model
plot_model(model, to_file='model_plot.png', show_shapes=True,
show_layer_names=True
```

5.2.3 Rectified Activations (and Batch Normalization)

To train a very deep network using gradient descent, a series of chained partial derivatives have to be calculated. The idea is that the error is propagated back from the output back to the input, updating all the weights in the process.

All the theory can be explained and derived in terms of the flux of the gradient going back from the output to the input. The real breakthrough of the theory was to recognize that the main obstacle to the training of a very deep network was the sigmoid function that was used as an activation function.

Specifically, the logistic function has almost zero derivatives away from the origin. When a neuron is in these conditions, the flow no longer propagates backward through it, and the weights of connections entering the neuron are no longer changed: the neuron dies.

Nowadays, neural networks are trained with ReLU function:

$$\sigma(x) = \mathrm{ReLU}(x) = \max(x, 0) \qquad (5.2)$$

The ReLU function is always positive and for positive numbers has a derivative exactly of one preserving the flux.

So using the ReLU activation, we can train very deep networks.

Alas, the nonlinearity and the discontinuity of the ReLU function also give a very jagged loss function with many local minima.

So to speed up training, we have to resort to regularization techniques or by adding explicit terms that limit the norm of weights or with specific methods of neural networks such as dropout, early stopping, and batch normalization.

We dwell briefly on batch normalization, and during the rest of the chapter, we will understand why this technique is so effective.

Batch normalization consists of normalizing the batches of samples on which the network must train layer by layer.

To clarify this concept, let us take the case of the loss function of a simple linear model. The best practice dictates that you normalize the samples before training the model because we make the loss function more regular.

In the case of a loss function with non-normalized features, the gradient descent tends to make many more steps. The single step has a jagged zigzag behavior that makes training more difficult.

Thus normalizing the samples before training the model facilitates the convergence of the model.

Batch normalization does precisely this but layer by layer. And it works like magic.

In the original paper, the authors argued that the merit of the method's success was the reduction of the covariate shift, but we will see later that batch normalization has another significant effect, the reduction of gradient shattering.

5.2.4   Saliency Maps

Armed with this technical knowledge, we can go on with the more proper XAI part.

In computer vision, it is common to speak of saliency maps, that is, maps that indicate which parts of an image capture the attention of the person who looks at them; in our case, the observer is replaced by a model that is examining them.

We have already dealt with an example of saliency map with occlusion analysis, now armed with the ability to use the internal workings of a network, we will talk about gradient-based saliency maps that are more accurate and faster to calculate than in the pure black-box approach.

The naïve idea on the description of the response of a neural network is to make a sensibility analysis or to evaluate how the response changes as the input changes. Following the article by Simonyan et al. (2013), we can calculate

$$\text{Sensibility} = \left(\frac{\Delta \text{Output}}{\Delta \text{Input}} \right)^2 \qquad (5.3)$$

This is equivalent to propagating information backward from the output class to the individual pixels of the image.

But the use of the ReLU as an activation function has the consequence that the DNNs as a whole are difficult to differentiate. They are locally linear in pieces, so their derivative is constant in pieces. The result of a simple analysis through the derivative will be very inaccurate and cannot discriminate between different classes (Fig. 5.8).

Input image   Backpropagation

Fig. 5.8 Directly backpropagating the output class probability to the input image pixels is very inaccurate and cannot discriminate between different classes

Several partial solutions to this problem have been created, and each of them uses brilliant and powerful ideas (we list a few).

5.3   Opening Deep Networks

5.3.1   Different Layer Explanation

In a computer vision neural network, the image is usually first processed by convolutional filters and then crosses the fully connected layers and a softmax to get the probabilities of the classes.

We have already said that convolutional filters have the remarkable property of being invariant by translation, so it is not important to know where a cat will be in the image; they will still react to its presence.

The fully connected layers, on the other hand, are excellent for decoding the output of the convolutional layers but completely break invariance due to translations.

From this, we can understand the usefulness of having methodologies that explain a neural network's working not starting from its last layer's output but starting from the output of any layer or even from disconnected sets of neurons or filters.

The next methods, among the most used, will take the input-output directly from the last convolutional layer.

5.3.2   CAM (Class Activation Maps) and Grad-CAM

The CAM methodology was one of the first to be devised and also one of the most widespread.

Class activation maps of Zhou et al. (2016) help us understand which regions of an image affect the output of a convolutional neural network.

The technique is based on a heatmap that highlights, as in the agnostic case, the pixels of the image that push a model to associate a certain class, a certain label, to the image.

It is noteworthy that the layers of a CNN behave in this case as unsupervised object detectors.

The implementation of the CAM technology is based on the properties of the global average pooling that were added after the last convolutional layer to decrease the size of the image and reduce the parameters so as to reduce overfitting.

The global average pooling layer works in the following way.

Each image class in the dataset is associated with an activation map, and the GAP layer calculates the average of each feature map.

We can see the representation in the image (Fig. 5.9).

The assumption of the CAM model is that the final score can always be expressed as a linear combination of the global pooled average of feature maps.

The CAM procedure is thus removing the last fully connected layers, applying a GAP to the last convolutional layer and training the weights from the reduced layer to the classes.

The linear combination allows us to have the final visualization.

This procedure, although efficient, has the drawback of changing the structure of the network and retraining it; furthermore, it can only be used by applying it only starting from a convolutional layer and therefore is not applicable in all architectures.

An evolution of the CAM is the Grad-CAM of Selvaraju et al. (2017). Grad-CAM does not retrain the network. Evaluate the weights of the linear combination, starting from the value of the gradient at the exit from the convolutional layer, and then apply a ReLU function to regularize it (Fig. 5.10).

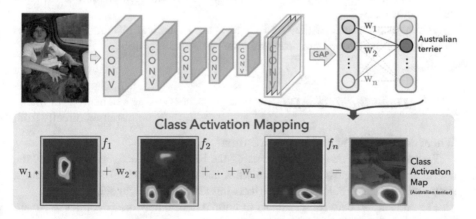

Fig. 5.9 The global average pooling gives the class activation map as a linear combination of features

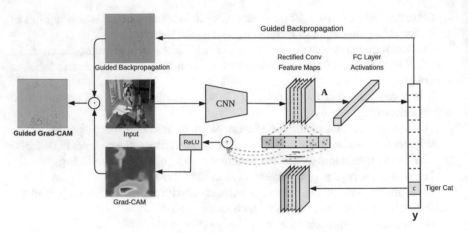

Fig. 5.10 Grad-Cam schema: visual explanations from deep networks via gradient-based localization

To obtain the heatmap, Grad-CAM calculates the gradient of y^c (the probability of class c) with respect to the feature map A of the convolutional layer.

These backpropagation gradients are averaged in the sense of the GAP to obtain the weights a_k^c – in formula (5.4) the summation represents the global average pooling, and the partial derivates are the gradients via backpropagation:

$$\alpha_k^c = \frac{1}{Z}\sum_i \sum_j \frac{\partial y^c}{\partial A_{ij}^k} \tag{5.4}$$

$$L_{\text{Grad–CAM}}^c = \text{ReLU}\left(\sum_k \alpha_k^c A^k\right) \tag{5.5}$$

The final contribution comes from an adjusted linear combination of the individual feature maps.

The main drawback of the Grad-CAM is the numerical instability that derives from the aforementioned gradient problems of neural networks.

5.3.3   DeepShap/DeepLift

Where do the gradient problems come from?

We have already mentioned that using the ReLU as an activation function, the neural network is almost always locally flat, and the gradient itself is discontinuous; shattered gradient problem further affects this situation (Balduzzi et al. 2017).

The correlation between gradients in normal neural networks declines exponentially with depth until a white noise pattern is obtained.

A partial solution to the shattered gradient problem is to "cure" the activation functions by ensuring that the gradient flow through them does not distort.

Currently, the reference method is that of Deep Learning Important Features or DeepLift (Shrikumar et al. 2017).

DeepLift is a method that decomposes the prediction of a single-pixel neural network. This is done by carrying out the backpropagation of the contribution of all neurons in the network for each input feature. DeepLift compares the activation of each neuron with its reference activation and evaluates the importance of each contribution, starting from this difference. DeepLift can reveal dependencies that may be hidden in other approaches; in fact, unlike other gradient-based criteria, it can also flow information through neurons with a zero gradient, a partial solution to the shattered gradient problem.

Unfortunately, to achieve all this, it is necessary to replace each activation function with that of the DeepLift, so the solution via DeepLift, unless you retrain the network, is a solution through a surrogate model.

Let's use the DeepLift implementation in the SHAP library for a practical example; taking a cue from the Keras tutorial, we train a model on the mnist dataset for the classification of the figures, and then we launch the deep explainer:

```
# DeepShap using DeepExplainer
# ...include code from https://github.com/keras-team/keras/
blob/master/examples/mnist_cnn.py

from __future__ import print_function
import keras
from keras.datasets import mnist
from keras.models import Sequential
from keras.layers import Dense, Dropout, Flatten
from keras.layers import Conv2D, MaxPooling2D
from keras import backend as K

batch_size = 128
num_classes = 10
epochs = 1

# input image dimensions
img_rows, img_cols = 28, 28

# the data, split between train and test sets
(x_train, y_train), (x_test, y_test) = mnist.load_data()

if K.image_data_format() == 'channels_first':
    x_train = x_train.reshape(x_train.shape[0], 1, img_rows,
    img_cols)
    x_test = x_test.reshape(x_test.shape[0], 1, img_rows,
    img_cols)
```

```
        input_shape = (1, img_rows, img_cols)
    else:
        x_train = x_train.reshape(x_train.shape[0], img_rows,
        img_cols, 1)
        x_test = x_test.reshape(x_test.shape[0], img_rows, img_
        cols, 1)
        input_shape = (img_rows, img_cols, 1)

    x_train = x_train.astype('float32')
    x_test = x_test.astype('float32')
    x_train /= 255
    x_test /= 255
    print('x_train shape:', x_train.shape)
    print(x_train.shape[0], 'train samples')
    print(x_test.shape[0], 'test samples')

    # convert class vectors to binary class matrices
    y_train = keras.utils.to_categorical(y_train, num_classes)
    y_test = keras.utils.to_categorical(y_test, num_classes)

    model = Sequential()
    model.add(Conv2D(32, kernel_size=(3, 3),
                      activation='relu',
                      input_shape=input_shape))
    model.add(Conv2D(64, (3, 3), activation='relu'))
    model.add(MaxPooling2D(pool_size=(2, 2)))
    #model.add(Dropout(0.25))
    model.add(Flatten())
    model.add(Dense(128, activation='relu'))
    model.add(Dropout(0.5))
    model.add(Dense(num_classes, activation='softmax'))

    model.compile(loss=keras.losses.categorical_crossentropy,
                  optimizer=keras.optimizers.Adadelta(),
                  metrics=['accuracy'])

    model.fit(x_train, y_train,
              batch_size=batch_size,
              epochs=epochs,
              verbose=1,
              validation_data=(x_test, y_test))
    score = model.evaluate(x_test, y_test, verbose=0)
    print('Test loss:', score[0])
    print('Test accuracy:', score[1])
```

After training we launch the deep explainer:

```
#DeepShap using DeepExplainer


import shap
import numpy as np

# select a set of background examples to take an expectation
over
background = x_train[np.random.choice(x_train.shape[0], 100,
replace=False)]

# explain predictions of the model on four images
e = shap.DeepExplainer(model, background)
# ...or pass tensors directly
# e = shap.DeepExplainer((model.layers[0].input, model.
layers[-1].output), background)
shap_values = e.shap_values(x_test[1:5])

# plot the feature attributions
shap.image_plot(shap_values, -x_test[1:5])
```

In Fig. 5.11 we can see all the pixels for or against the number's attribution to some class. So in the first image, we see a circle formed by blue pixels that are the missing pixels for the zero class' proper attribution. In the third image, where the two is correctly classified in the two-class, the picture is full of red pixels.

5.4   A Critic of Saliency Methods

5.4.1   What the Network Sees

So far, through salience methods, we have seen how to highlight the features of an image that most influence the output of the model.

However, this way of proceeding has an intrinsic limit, which is clearly illustrated in a sentence by Cynthia Rudin (2019):

"Saliency maps are often considered to be explanatory. Saliency maps can be useful to determine what part of the image is being omitted by the classifier, but this leaves out all information about how relevant information is being used. Knowing where the network is looking within the image does not tell the user what it is doing with that part of the image."

We can say salience tells us what the network sees, not what the network thinks.

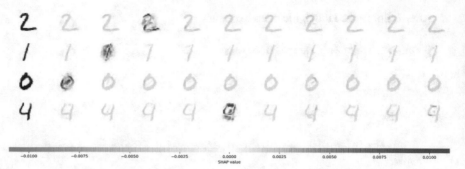

Fig. 5.11 DeepShap gives an attribution score for the class coloring the pixels in the images

That is, it gives us feature importance but in no way illustrates the process by which that feature is analyzed.

We will try to answer this limit in this section by citing a beautiful work in terms of gradient flow analysis and in the next section by unsupervised methods.

5.4.2   Explainability Batch Normalizing Layer by Layer

A fascinating article by Chen et al. (2020) allows us to return to the often underestimated importance of batch normalization. As we mentioned in the past paragraphs, batch normalization was introduced to prevent the input of neurons from being too large and facilitate the convergence of the hanging and regularizing the loss function making it more spherical near the minimum.

Rudin's work allows us, once again, to show how a methodology designed to accelerate the learning process has its roots in Explainable AI.

That is, how each request to have an Explainable AI method at a deeper level leads to more robust and effective learning.

Which Explainable AI request does batch normalization respond to?

Let's think of the classic example of CIFAR10: classifying images in ten different classes.

Rudin wonders if it is possible to follow the flow of the "concept" of class between layer and layer and if it is possible to build a new type of neural network that layer by layer manages to keep concepts (classes) as separate as possible.

To do this, all he does is use a batch normalization procedure and reorganize the information in each layer (appropriately applying linear transformations).

Let's not go deep into the math involved. We are satisfied with the idea (Fig. 5.12).

Thus it trains a special type of network that is able to follow the flow of class information from input to output.

This work not only shows us how neural networks can be trained with the requirement of interpretability but also tells us that batch normalization is an indispensable requirement for this interpretability. We can even think of explainability as a regularizing procedure.

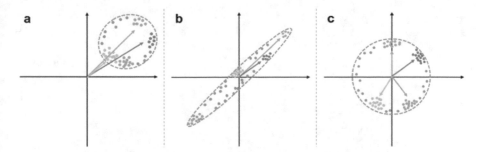

Fig. 5.12 The data distributions in the latent space. (**a**) The data are not mean centered; (**b**) the batch normalized but not decorrelated; (**c**) the data are fully decorrelated (whitened). In (**c**) the unit vectors can represent concepts

Take this paragraph as a glimpse of the future: we don't know what the neural networks of tomorrow will be like, but we bet they will be interpretable!

5.5   Unsupervised Methods

In this section, we will see how unsupervised methods can help us analyze what the network has learned.

5.5.1   Unsupervised Dimensional Reduction

Non-supervised methods are a field in active development.

In this section, we will limit ourselves to a classic use of them: dimensional reduction. Dimensional reduction is a very powerful method to control the curse of dimensionality by reducing the number of features of the samples, and at the same time, it allows us to draw representations of our datasets in low-dimensional spaces.

Let's take a quick example using two of the best-known techniques:

Principal component analysis (PCA) and t-SNE. PCA is a linear transformation that maps our dataset in one that produces new features along the dimensions of the largest variance.

t-SNE instead is a nonlinear method; it leverages Kullback-Leibler divergence so that the statistical properties of the samples in the high-dimensional space and the low-dimensional one are the same.

Using `sklearn` we will map the 64-dimensional dataset of small images of digits (8 × 8 pixels each) to a 2-dimensional space (Fig. 5.13).

```
import numpy as np
import matplotlib.pyplot as plt
from sklearn import manifold, decomposition, datasets
```

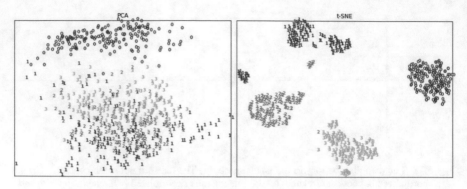

Fig. 5.13 PCA and t-SNE dimensional reduction reducing from 8 × 8 = 64 dimensions to 2. t-SNE clustered MINST digits more effectively

```
digits = datasets.load_digits(n_class=4)
X = digits.data
y = digits.target
n_samples, n_features = X.shape


def plot_lowdim(X, title=None):
    x_min, x_max = np.min(X, 0), np.max(X, 0)
    X = (X - x_min) / (x_max - x_min)

    plt.figure(figsize=(8,6))
    ax = plt.subplot(111)
    for i in range(X.shape[0]):
        plt.text(X[i, 0], X[i, 1], str(y[i]),
                 color=plt.cm.Set1(y[i] / 10.),
                 fontdict={'weight': 'bold', 'size': 9})
    plt.xticks([]), plt.yticks([])
    if title is not None:
        plt.title(title)


#--------------------------------------------------------------------
# PCA

X_pca  =   decomposition.TruncatedSVD(n_components=2).fit_trans
form(X)
plot_lowdim(X_pca,"PCA")

# t-SNE
tsne = manifold.TSNE(n_components=2, init='pca', random_
state=0)
X_tsne = tsne.fit_transform(X)
```

```
plot_lowdim(X_tsne,"t-SNE")

plt.show()
```

We can see from the images that similar figures are grouped, and intuitively we can get an idea of how much t-SNE does a better job by leaving a larger space between the clusters.

5.5.2   Dimensional Reduction of Convolutional Filters

To help understand what activation atlases are, we illustrate a work by Karpathy (2014).

Karpathy uses t-SNE as an unsupervised method of investigating neural networks.

He extracts the 4096-dimensional output from the seventh layer of the famous AlexNet (Krizhevsky et al. 2017) and feeds it to t-SNE (Fig. 5.14).

Now the output of t-SNE can be seen as a possibly unique hash code of the image seen by the neural network. In his work, Karpathy uses this code to have a two-dimensional representation of the connected layer with relative clustering.

And by showing images at NN, we see how they are mapped in 2D space (Fig. 5.15).

We will say that, by using the dimensional reduction algorithms, we can create an atlas of the images shown on a CNN network.

This representation allows us to understand how the network thinks without detailing it. The network shows nearby images that it considers similar, and this allows us both to isolate errors in the dataset (e.g., a number 1 in the same cluster of numbers 5) and to empirically evaluate the effectiveness of learning. It is heartening to see in the previous figure that all the animals are combined, but this could be the effect of some adversarial feature rather than an actual semantic division.

To know what the network really thinks, we need a more refined tool.

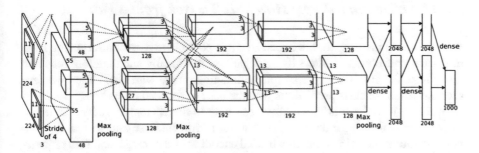

Fig. 5.14 AlexNet architecture. A succession of convolutional filters and Max pooling reductions to other convolutional filters and at the end of the network two fully connected dense layers outputting to the output 1000 class vectors with a softmax function

Fig. 5.15 An atlas of images using the dimensional reduction technique (Karpathy 2014)

5.5.3 Activation Atlases: How to Tell a Wok from a Pan

With the CAM method, we have seen how it is possible to extract information from individual convolutional filters by aggregating them locally and therefore reducing their dimensionality through a simple arithmetic average with GAP. Then we have seen how to make class predictions with the base of reduced filters.

With the t-SNE method, we have seen how to dimensionally reduce an entire image by transforming it through the network, taking the filter values of all the convolution filters and using them as a hashed encoding of the image, and then organizing the images of the train set into a two-dimensional grid. The two-dimensional grid showed us the similarity between images with respect to some metric learned from the neural network in class recognition.

Now let's go to the next step.

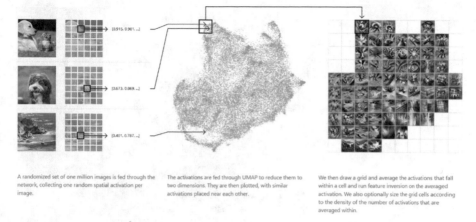

A randomized set of one million images is fed through the network, collecting one random spatial activation per image.

The activations are fed through UMAP to reduce them to two dimensions. They are then plotted, with similar activations placed near each other.

We then draw a grid and average the activations that fall within a cell and run feature inversion on the averaged activation. We also optionally size the grid cells according to the density of the number of activations that are averaged within.

Fig. 5.16 One million images are given to the network, one activation function per image. Activations are dimensionally reduced to two dimensions. Similar activations close together

In recent joint work, DeepMind and OpenAi, Carter et al. (2019) have used size reduction to classify not the training set images but directly the convolutional filters.

That is, they used size reduction not so much to see how CNN classifies things rather the density of the filters used to recognize things: an atlas of activation maps (Fig. 5.16).

Activation atlases not only allow us to see what the network thinks but also to more carefully sampling the activation functions that will be used most frequently.

After calculating which activation functions are close to each other, the functions in the same cell are averaged. Of the mediated activation functions, it searches which images came from with a feature visualization technique.

The result from a theoretical point of view is amazing, but let's also see a practical application.

Suppose that as a data scientist you may be asked to find some erroneous training in a model seeking counterfactual answers.

The visualization atlas technique can be a very powerful one because it can revert the search to the way the model looks at the images.

Say we want to search what the network thinks seeing a wok (or a frying pan) or, more precisely, what maximally excites CAM to output the answer.

In the meantime, we can do a search of which are the classes of activations that respond the most when we show images of wok and frying pan to the network (Fig. 5.17).

Now we can see the activation atlas shows us many examples of a wok, and in some of them, there are noodles.

If we think of our model as a discriminator, say of wok and frying pans, we can see that the noodles are absent in frying pans' visualization atlas (Fig. 5.18).

Perhaps we have found a counterfactual feature, and in fact, by overlaying a sticker with noodles on the photos of a frying pan, we can make the neural network believe that it sees woks.

Alas, our neural network is actually a wok discriminator.

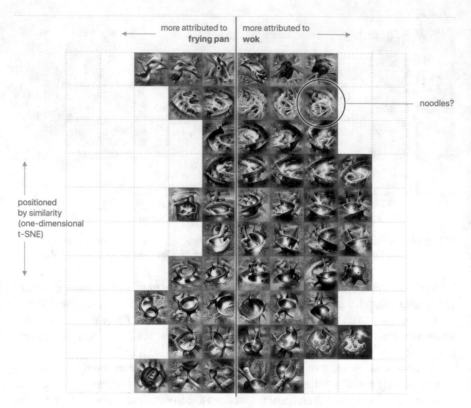

Fig. 5.17 Is it a frying pan or a wok? These are the most excited filters at the sight of frying pan and wok

5.6 Summary

- In the context of the agnostic approach, we have introduced a parallelism between robust learning and explainability.
- We introduced adversarial features and augmentations.
- We have made the parallelism between occlusions used as augmentations and an agnostic XAI method.
- We introduced neural networks by presenting their structure, equations, and convergence problems.
- We answered the question about the need for Deep Neural Networks and introduced batch normalization.
- We opened the neural networks by introducing the concept of salience the explanation starting from different layers and the CAM (class activations maps) and Grad-CAM methods.
- We have introduced DeepLift as a solution to the gradient shattering problem.
- We carried out a critique of salience methods by presenting a work on interpretability via batch normalization.

1.	frying pan	76.5%
2.	wok	15.8%
3.	stove	5.4%
4.	spatula	1.0%
5.	Dutch oven	0.5%
6.	mixing bowl	0.2%

1.	wok	63.2%
2.	frying pan	35.1%
3.	spatula	0.6%
4.	hot pot	0.5%
5.	mixing bowl	0.1%
6.	stove	0.1%

Fig. 5.18 We have a perfect noodle discriminator

- We introduced unsupervised methods that use size reduction for the semantic representation of what the network has learned.
- We have introduced the activation atlases capable of showing us how the web sees the world.

References

Balduzzi, D., Frean, M., Leary, L., Lewis, J. P., Ma, K. W. D., & McWilliams, B. (2017). The shattered gradients problem: If resnets are the answer, then what is the question? *arXiv preprint arXiv:1702.08591*.

Bottou, L. (2019). *Learning representations using causal invariance*. Institute for Advanced Studies talk at https://www.youtube.com/watch?v=yFXPU2lMNdk&t=862s.

Carter, S., Armstrong, Z., Schubert, L., Johnson, I., & Olah, C. (2019). Activation atlas. *Distill, 4*(3), e15.

Chen, Z., Bei, Y., & Rudin, C. (2020). Concept whitening for interpretable image recognition. *arXiv preprint arXiv:2002.01650*.

Cybenko, G. (1989). Approximation by superpositions of a sigmoidal function. *Mathematics of Control, Signals, and Systems, 2*(4), 303–314.

Hornik, K. (1991). Approximation capabilities of multilayer feedforward networks. *Neural Networks, 4*(2), 251–257.

Karpathy, A. (2014). *t-SNE visualization of CNN codes*. https://cs.stanford.edu/people/karpathy/cnnembed/.

Krizhevsky, A., Sutskever, I., & Hinton, G. E. (2017). Imagenet classification with deep convolutional neural networks. *Communications of the ACM, 60*(6), 84–90.

Mhaskar, H., Liao, Q., & Poggio, T. (2019). Learning functions: When is deep better than shallow. *arXiv:1603.00988*. https://arxiv.org/abs/1603.00988.

Nakkiran, P., Kaplun, G., Bansal, Y., Yang, T., Barak, B., & Sutskever, I. (2019). Deep double descent: Where bigger models and more data hurt. *arXiv preprint arXiv:1912.02292*.

Rudin, C. (2019). Stop explaining black box machine learning models for high stakes decisions and use interpretable models instead. *Nature Machine Intelligence, 1*(5), 206–215.

Selvaraju, R. R., Cogswell, M., Das, A., Vedantam, R., Parikh, D., & Batra, D. (2017). Grad-CAM: Visual explanations from deep networks via gradient-based localization. In *Proceedings of the IEEE international conference on computer vision* (pp. 618–626).

Shrikumar, A., Greenside, P., & Kundaje, A. (2017). Learning important features through propagating activation differences. *arXiv preprint arXiv:1704.02685*.

Simonyan, K., Vedaldi, A., & Zisserman, A. (2013). Deep inside convolutional networks: Visualising image classification models and saliency maps. *arXiv preprint arXiv:1312.6034*.

Vapnik, V. (2000). *The nature of statistical learning theory*. New York: Springer.

Zeiler, M. (2013). *Visualizing and understanding convolutional networks*. https://arxiv.org/abs/1311.2901.

Zhou, B., Khosla, A., Lapedriza, A., Oliva, A., & Torralba, A. (2016). Learning deep features for discriminative localization. In *Proceedings of the IEEE conference on computer vision and pattern recognition* (pp. 2921–2929).

Chapter 6
Making Science with Machine Learning and XAI

"The hardest part of research is always to find a question that's big enough that it's worth answering, but little enough that you actually can answer it."
—Edward Witten

This chapter covers:

- How to do physics models with ML and XAI?
- Do we need causation to make science?
- How to effectively use ML and XAI in science?

At the very beginning of this book, we tried to clarify the difference between the terms interpretability and explainability. In that context, we said that *interpretability is the possibility of understanding the mechanics of a Machine Learning model, but this might not be enough to answer "Why" questions – questions about the causes of a specific event.* We also provided in Table 1.1 of Chap. 1 (don't worry to look at it now, we will start again from this table in the following) a set of operational criteria based on question to distinguish between interpretability as a lighter form of explainability. As we saw, explainability is able to answer questions about what happens in case of new data, "What if I do x, does it affect the probability of y?" and counterfactual cases to know what would have changed if some features (or values) would not have occurred. Explainability is a theory that deals also with unobserved facts toward a global theory, while interpretability is limited to make sense of what is already present and evident. The question is: Why are you getting back to this point in this chapter about making science with ML? The answer, long story short, is that explainability is exactly what we need to climb "the ladder of causation" (we will talk about it in a while). We will use XAI in the domain of "knowledge discovery" with a specific focus on scientific knowledge. To recall what we already discussed:

© The Author(s), under exclusive license to Springer Nature Switzerland AG 2021
L. Gianfagna, A. Di Cecco, *Explainable AI with Python*,
https://doi.org/10.1007/978-3-030-68640-6_6

Knowledge discovery is the most complex application to comment, being related to situations in which ML models are not just used to make predictions but to increase the understanding and knowledge of a specific process, event, or system. The extreme case that we will discuss further in the book is the adoption of ML models to gain scientific knowledge in which prediction is not enough without also providing explanations and causal relations.

The main objective of this chapter is to touch with hands how to use ML + XAI to get knowledge and study a real physical system beyond the usual scope of ML that is just to make predictions. We will use this scenario to clarify all the limitations, opportunities, and challenges of exploiting ML to make science.

6.1   Scientific Method in the Age of Data

How does Google rank the pages to propose you the best one for your search? Or equivalently, how does Google match the ads with user's preferences? From our purposes, the answer is all in the words of Peter Norvig (Google's research director): "All models are wrong, and increasingly you can succeed without them."

Google's success is not based on any "understanding" of the content of the pages. There is no semantic or causal analysis; it is just a complex algorithm based on the relative number of links that set the ranking. The news is that this kind of approach is now being adopted in science and might replace the classical scientific method. To understand whatever phenomenon, scientists rely on model building: they try to narrow down the essential variables that affect the outcome, build an approximated model based on these variables, and test the model with experiments.

The iterative process is (1) make hypothesis, (2) build a model, and (3) do experiments to test the model. The age of data may replace this with a new mantra: correlation is enough, we don't need to build models, as far as our objective is to predict an outcome, just feed the problem Machine Learning system with tons of data for the learning phase, and then you will have your results.

Despite the apparent oversimplification of the two alternatives, there is a lot to think about in such a state of things. On one side, science, and in particular physics, is running into fields where experiments are not easy or not possible at all (think about cosmology), so that it could make sense to rely on the effective prediction of ML instead on beautiful models but driven only by the "beauty" of the related mathematics. On the other side, the question about the real understanding that we may get just from a DNN applied to physics instead of having a model remains open. And this question is the one we want to answer in this chapter leveraging XAI.

The real case scenario that we will study to discuss and answer this question is to predict the position of a one-dimensional damped pendulum at different times. This is a straightforward physical system that can be easily solved with basic knowledge of physics and math. The observations are a set of couples in which we record the position of the pendulum x_i at different times t_i. The question for the ML system after training is: "Where is the pendulum at time t_k?"; but as you can easily guess,

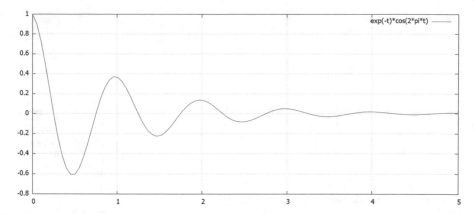

Fig. 6.1 Damped oscillation

this is not enough to say that we get full knowledge of the physics of the pendulum. We will see how to use XAI to achieve such an understanding.

Before getting into details, we need to make some further comments about this scenario. This should be new for our reader that is assumed to have basic in ML, but we want to emphasize the differences with what we did so far. The applications we worked out in this book deal with supervised and unsupervised ML models, but we never met time series like in this case.

Here, as we said, we have a damped pendulum, and we want to predict the x position at a time t (Fig. 6.1).

In this case, we don't have the usual input and output features like in the previous examples (e.g., we predicted and understood the most important features to assign the best player award for a football match in Chap. 4 based on the match metrics like the number of goals scored and so on). Time series needs to be re-framed as a supervised learning dataset using "feature engineering" to construct the inputs that will be used to make predictions (in this case, it would be the position of the pendulum).

Basically, a generic time series appears like this (Table 6.1):

And we need to transform this couple (time, value) to something like this to fit the usual supervised learning models (Table 6.2).

The most common classes of features that are created from the data series are:

Date time features: these are the timestamps for each observation, for example, they may be used to discover recurrent patterns and cycles of the target variable.

Lag features: these are values at a previous time. The underlying assumption is that the value of the target at time t is affected by the value at the previous time step. The past values are known as lags.

Window features: these are aggregated statistical vales of the target variable over a fixed window of time. This method is also known as the rolling window method because the timeframe to calculate the statistical values is different for each data point (e.g., rolling average).

Table 6.1 Time series

Time	Value
t1	Value1
t2	Value2
t3	Value3

Table 6.2 Supervised learning

Feature	Value
f1	Output1
f2	Output2
f3	Output3

Table 6.3 From time series to supervised learning

Transform a time series to a supervised learning						
Supervised		Time series		From time series to supervised		
x	Y	Time	Measure	Time	Lag1 as x	Y
5	1	1	1	1	?	1
6	0	2	0	2	1	0
9	1	3	1	3	0	1
8	0	4	0	4	1	0
9	1	5	?	5	0	?

Domain-specific features: these are at the foundation of feature engineering; the knowledge of the domain and the data guide the choice of the best features for the model.

So, the game is to pass from the time series to supervised case as in Table 6.3.

But having the time series adapted for a supervised learning is not enough for our goals. At this point we have the possibility to use the full supervised learning machinery to do predictions, but how to learn about the physics of the damped pendulum?

The obvious answer that you should have, based on our journey so far, is that XAI is exactly for this: to extract explanations from the ML model that has been built to predict the pendulum position after the proper transformation of the time series. And the explanations in this case should allow us to learn the physics of the pendulum. But unfortunately, this is not the case – at least in general terms, the techniques we explained so far are not enough to get scientific knowledge we are searching for.

We don't know how many features may be generated by our feature engineering of the time series, but the XAI methods we have would generate explanations about the most important features, rank them, and in case answer specific question for a specific data point (as we saw with SHAP), but this is not enough. This level is what we called "interpretability," but here we need to answer questions in the domain of knowledge discovery: discover the causal relations, and answer questions on unseen data.

To touch with hands what we are saying, the point is that we will never get the physics model of the pendulum that relies just on spring constant and damping factor to get the full understanding of this system. XAI methods we learned won't identify these two physical variables as the ones needed to solve the physics of the damped pendulum.

To get to this level we need a different approach that is the core of this chapter. But before getting into this, we need to climb "the ladder of causation" in order to better understand:

1. The limitations of the XAI methods we presented if applied to knowledge discovery.
2. Clarify, based on the point 1 above, once and for all what we mean by explainability in comparison with interpretability. At this point we have the skills, and we can get into details of this with the real case scenario of the damped pendulum.

6.2 Ladder of Causation

In Chap. 1 we used Table 1.1 to distinguish between interpretability and explainability; we put it here again for simplicity and as core of our reasoning (Table 6.4):

Getting back again to Gilpin's words: "We take the stance that Interpretability alone is insufficient. For humans to trust black-box methods, we need Explainability – models that can summarize the reasons for neural network behavior, gain the trust of users, or produce insights about the causes of their decisions" (Gilpin et al. 2018). As we discussed, the table uses two different sets of questions in order to distinguish between interpretability and explainability; there are questions that cannot be

Table 6.4 Difference between interpretability and explainability in terms of the questions to answer for the two different scopes

Question	Interpretability	Explainability
Which are the most important features that are adopted to generate the prediction or classification?	✓	✓
How much does the output depend on the input? How sensitive is the output on small changes in the input?	✓	✓
Is the model relying on a good range of data to select the most important features?	✓	✓
What are the criteria adopted to come across the decision?	✓	✓
How would the output change if we put different values in a feature not present in the data?	✗	✓
What would happen to the output if some feature or data had not occurred?	✗	✓

answered in the domain of interpretability. To make sense of these questions that, as we will see better in this chapter, are fundamental for scientific knowledge, we will follow the seminal work of Pearl & Makenzie (2019) on causality.

The core of Pearl's line of research can be summarized with a picture of the ladder of causation (Fig. 6.2):

There are three different types of cognitive abilities that are needed to climb the ladder of causation: seeing, doing, and imaging. The highest level, that is, imagination, is what allowed for the impressive progress that humans did from our *Homo sapiens* ancestors until our age of data. Already *Homo sapiens* were using imagination to think about situations that were just potential, planning for hunt meant to plan for "unseen" things that could happen.

At the base of the ladder, we have association that is connected to the activities of seeing and observing. What we do here is to search for patterns, regularities in what we see or in a huge amount of data. The goal is to find correlations that may help us to do predictions so that observing one event may change the probability of observing the other.

The classical examples that are specific to this domain come from marketing. Imagine being a marketing director that wants to understand how likely it is that a customer that bought an iPhone also bought an iPad. The answer will be based on collecting the data, segmenting the customers, and focusing on the group of people that bought an iPhone. Then we may compute the proportion of these people that get also and iPad – this is just a way to compute the conditional probability of an event given another based on existing data: P(iPad|iPhone).

When we explained the XAI methods of linear regression or logistic regression, we just did this: search for correlations in the data that are not necessarily linked to causal relations. As in this case, we cannot say that buying an iPad is the cause of buying an iPhone or vice versa, but for our marketing purposes, it is enough to know the degree of association between the two events. The predictions coming from the first rung of the ladder are based on passive observations, and the related XAI methods answer the questions placed in the interpretability column. Moving from the marketing scenario to our damped pendulum, we can predict the position of the pendulum through the feature engineering of the time series, and we may narrow down the most important features with XAI but without any knowledge discovery. Using Searle's words: "Good predictions need not have good explanations. The owl can be a good hunter without understanding why the rat always goes from point A to point B. Some readers may be surprised to see that I have placed present-day learning machines squarely on rung one of the Ladder of Causation, sharing the wisdom of an owl."

But this will be clear progressing with an investigation of the second and third rung of the ladder, and indeed we will see how to tackle this state of things and do science with the recent progress of ML and XAI.

Climbing to rung two, we enter the domain of doing instead of seeing; this is different from the first rung in which we just did associations on existing data. In this case, we want to know how predictions would change if we do a specific action.

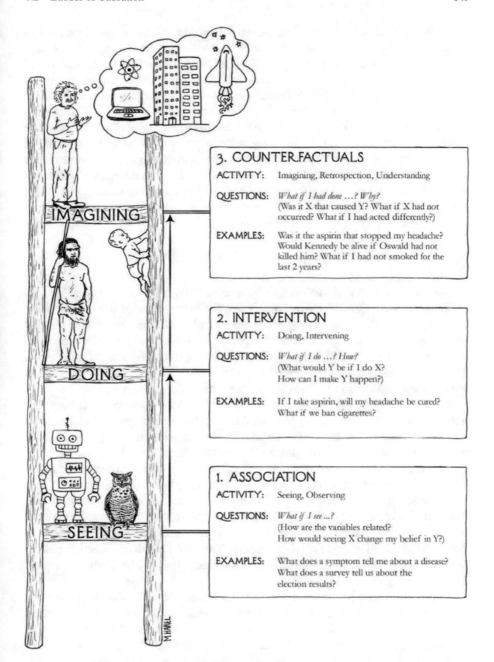

Fig. 6.2 Ladder of causation (Pearl & Makenzie 2019)

The difference is a bit tricky. Remember the marketing case of the first rung in which we investigate the selling of iPads conditioned on the selling of iPhones. A typical question of rung two would be: What if I double the price of iPhones? Would the relation with the sales of iPhones change? To answer such types of questions, we cannot easily rely on the collected data. Albeit we may find in our huge database the data referring to the case of iPads with doubled price, this is totally different from an intervention on the current market in which we double the price of iPhones. The existing data we may find would likely refer to a totally different background in which the price was double for different reasons (short of supply?). But here we are not asking for the probability of selling iPads conditioned on the sales of iPhone of a certain price, we are asking for the probability of selling iPads conditioned on the sales of iPhones with the intervention of doubling the price of iPhones. And in general, as shown by Pearl with his casual diagrams:

$$P\left(iPad|iPhone\right) \text{ is different from } P\left(iPad|do\left(iPhone\right)\right)$$

How can we move to this rung of intervention? What is usually done is to do experiments in controlled environments like the big companies usually do: think about Amazon, changing or suggesting items with a different price to a selected set of customers to see how it goes. If experiments are not possible, the alternative is to do a kind of "causal model" of the customer, including market conditions. This causal model is the only option to go from observational data of rung one to answer the question on rung two that assumes an explicit intervention.

Despite the wording that is not so common, intervention is something that belongs to our daily lives. Every time we decide to take a medicine for headache, we are doing a rung two intervention in which we are implicitly modeling a causal relation between the medicine we took and headache. Our belief is based on the "controlled" experiments that showed that the medicine is expected to remove headache. But there is another last step to climb the ladder to rung three, that is, the one of pure scientific knowledge. Rung three is the domain of counterfactual questions like, "What if I had acted differently?" This means to change something that already happened in the past, change the course of what is previous in time that is different from doing an intervention, and see what happens. The world in which an action has not been performed does not exist because it has already passed. That's why the typical activity of rung three is imaging.

What would have happened if I had not taken medicine? The data to answer this question do not exist by definition. But albeit weird, this is exactly what we need, and we use to do science. The laws of physics can be thought of as counterfactual assertions. Let's think about again our damped pendulum. As soon as you understand the physics of the systems, you will know that there are just two variables that control the pendulum's dynamics: the spring constant and the damping factor (we will details this later). Relying on these two constants, we have the full functional relationship to predict the position at any time t. How did we get this causal model? Starting from rung one of the associations, climbing to rung two in which physicists may experiment with different pendulums, in different conditions and different values of spring constant and damping factor, so to reach rung three that allows answering questions for any possible instances of the damped pendulum. Having the causal model means that there is no difference between the existing world in which you have the real observations of the time series for the pendulum and whatever imaginary world with hypothetical values for the same constants.

Back to the marketing, being in rung three would allow answering questions like, "What is the probability that a customer who bought the iPad would still have bought it in case of a double price?" In the real world, this didn't happen, but we have a model to answer this.

Rung three is the proper level of causal modeling that is needed to do science. And from our point of view, this is explainability as a stronger version of interpretability that, with the XAI methods we saw, helps in answering questions up to rung two. But we will see in the next section how to touch with hands XAI methods that will allow us to work with ML to acquire true scientific knowledge; the approach, based on what we said so far, will be different from just searching for the most important features, but it will deal with building a proper representation of the physics of the damped pendulum in order to allow ML to answer questions belonging to rung three.

Before jumping to the next section, we also propose a short optional content that you are free to skip (not fundamental to get the core of the rest), but it could be interesting to further detail the problem of causal knowledge. We will talk about Pearl's mini Turing test.

The question behind mini Turing test is the following: How can a ML system represent causal knowledge in order to answer rung three questions on a simple story? As you may see, this is a variation of original Turing test and Searle's Chinese Room we discussed in Chap. 1. Let's look at it through an example, presented by Pearl, that is about a firing squad in which there is a dramatic chain of events in which a Court Order is received by a Captain who passed it to two soldiers whose task is to fire to Prisoner D if order is received. The soldiers will fire only if the command is received, and if they fire (one of them is enough) Prisoner D is dead. We can suppose that our ML system is trained on data that records different sets of the five variables with different states.

Let's ask questions as examples from the different rungs of the ladder:

Rung 1: If the D = true (prisoner is dead), does it mean that that CO = true (the order has been given)?
This is pretty trivial, and even without any deep dive of the causal diagram, the answer is yes. It would be enough for our ML to track the associations on the five variables to make the proper prediction (Fig. 6.3).

Rung 2: What if soldier A decides to shoot without the order? This is a tricky question. Let's look at how our causal diagram would change (Fig. 6.4).

(continued)

(continued)

The intervention removed the link between CO and A; A is true without CO being necessarily true; prisoner is dead whatever CO is because A fired. And this is exactly the difference between "seeing" and "doing," climbing from rung one to rung two. B is untouched by the intervention, before the intervention A and B were necessarily coupled, both depending on CO. Our intervention allowed D = true with A = true and B = CO = false. Assume that the ML system doesn't have the causal representation of this story; it would not pass the mini-Turing test, because ML would be trained on thousands of records of executions, but "normally" all the variables would be all true or all false. ML system would not have any way to answer the question about what would happened to the prisoner in case we persuade A not to fire without having a causal representation of the relations among the events. And a rung three question would follow the same logic:

Rung 3: Suppose we have D = true, prisoner is dead. What would have happened if A had decided not to shoot? That is to compare the real world with a fictitious world in which A didn't shoot. And the causal diagram helps us as well making clear the state of the CO = true, B = true, and D = true (Fig. 6.5).

The prisoner would have died also in the imaginary world.

Fig. 6.3 Causal diagram for the firing squad example (Pearl & Makenzie 2019)

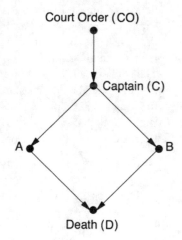

Fig. 6.4 Case of intervention, the link between C and A is removed, A is set to true whatever C is (Pearl & Makenzie 2019)

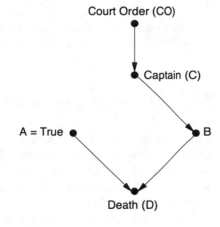

Fig. 6.5 Counterfactual reasoning, D is set to dead; what would have happened if A had not fired? (Pearl & Makenzie 2019)

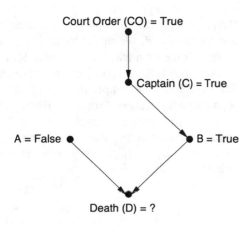

6.3 Discovering Physics Concepts with ML and XAI

Based on what we discussed in the previous sections, we are now ready to climb our ladder of causation to get the physics of a damped pendulum with Machine Learning and XAI.

To recap the flow, the usual approach of ML to a problem like the damped pendulum would be to set a neural network to train on the time series after a bit of feature engineering. This would work fine to forecast the position of the pendulum at different times, and we may use the XAI methods we learned in the previous chapters to get some insight on the most important features. But this would not allow us to climb our ladder of causation up to rung three, that is, the domain of counterfactuals and knowledge discovery, which in this case is the physics of the damped pendulum. We will show in this section how to tackle this challenge; we need to change our perspective and rely on a different type of neural networks to focus more on the proper "representation" than on just the predictions. To do this we will adopt the so-called autoencoders that are well known in the ML field, but here we will look at them for our specific knowledge discovery purpose.

6.3.1 The Magic of Autoencoders

Current architectures for artificial neural networks in Machine Learnings are built of huge number of layers and thousands of nodes connected in different ways. The different topologies of the connections produce the specific behavior of the networks and shape their ability to perform specific computational tasks.

A feedforward NN may look like this (Fig. 6.6):

There is an input layer that transmits the values to internal layers that do the computation to produce the output. As we know, the training of the NN is basically the process to find the right weights of each node in the internal layers to produce the proper output. We recall this basic info about a general feedforward neural network to compare it with the logic of autoencoders.

Autoencoders are a specific type of NNs that are not used to make predictions but just to reproduce as output the input that is provided. This could appear as a silly game; let's think about a topology like the one in Fig. 6.7:

It is a fully connected model of NN, and it is pretty obvious that if our goal is just to reproduce the output, the training would generate a solution like the one below, a kind of identity matrix that just propagate the input to the output without any added value (Fig. 6.8).

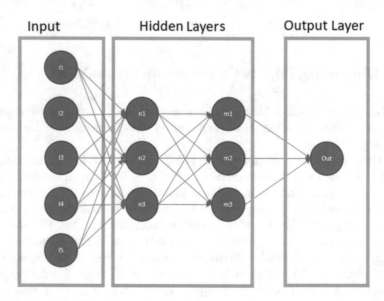

Fig. 6.6 A generic feedforward neural network

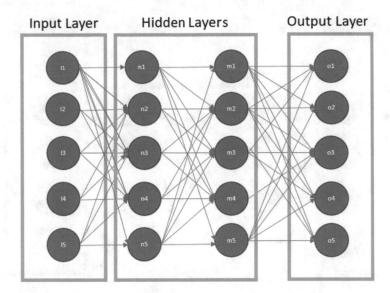

Fig. 6.7 Fully connected neural network

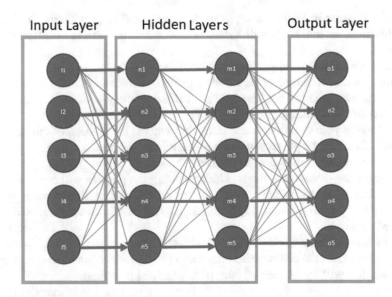

Fig. 6.8 Trivial Neural Network topology

But imagine now to reduce the number of nodes in the hidden layers (Fig. 6.9); in this way the autoencoder cannot just "pass" the input to the output, but it is forced to do some kind of compression to achieve the goal of reproducing the input.

The internal layers have a reduced number of nodes, and this makes the autoencoder to represent the information in a compressed form. Basically, the autoencoder follows a two-step process: an encoding phase in which the input is reduced and a

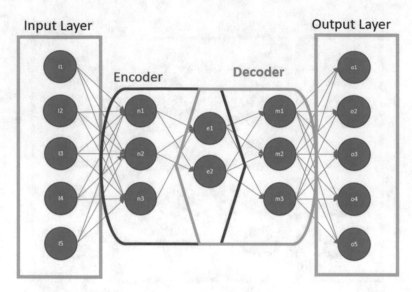

Fig. 6.9 Autoencoder topology

decoding process in which the input is "reconstructed" to provide the output. The loss function, in this case, will measure how much the output is different from the input.

The more general architecture looks like this:

Note that the decoder has the same (mirrored) architecture as the encoder part; this is very common but not a must (Fig. 6.10). Assume you have as input a hand-written number; the flow appears like in Fig. 6.11.

I can guess your question at this point: but what is the relation of this stuff with our goal of getting the physics of damped pendulum? We need a little more patience, but we may anticipate that the answer is the "Code" layer that you see in the middle of the encoding-decoding process. Understanding the physics will mean using the autoencoder to find the most "compact" representation of the physical system, which is to find the relevant physical variables to have a complete model of the damped pendulum.

What we call here "Code," also called latent space representation, that is, the layer that contains the compressed information to reproduce the input (number 4 in the example), will contain the physics of the damped pendulum.

The only additional step we need to do before putting hands into the specific problem of the damped pendulum is to remark that we will use a slight variation of the autoencoder names: variational autoencoders. For our purposes, it is enough to get that the main feature of VAEs compared to AE is that VAEs don't just learn a function to get a compressed form of the input, but it is a generative model. VAEs learn a function that is also able to generate variations around the "model" learned by the input; that is what we are searching for: we don't want to solve just the case of our specific damped pendulum but learn the physics of whatever damped pendulum to answer questions on variations.

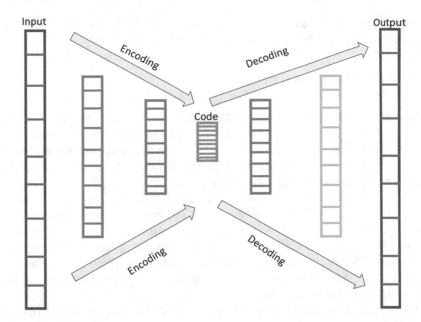

Fig. 6.10 Autoencoder process

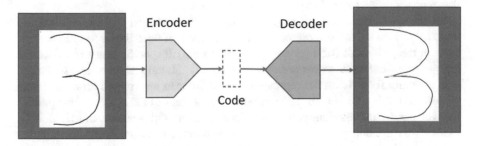

Fig. 6.11 Number recognition with autoencoder

6.3.2 Discover the Physics of Damped Pendulum with ML and XAI

After this bunch of theory, let's clearly formulate again the practical scenario we want to address. As XAI experts we are asked to get knowledge on the physics of a system with ML. We discover that just doing predictions and using common XAI techniques are not enough because we want to answer questions in rung three of our ladder. The physical system we will use to explain the approach is a damped pendulum based on the work of Iten et al. (2020). No need to remark that this is a very simple physical system, but despite this, it will allow us to show the direction and the techniques to use for our purpose.

We have a time series x_t where x_{ti} represents the position of the damped pendulum at the time t_i. After the training, we want to achieve two goals with our ML system:

1. Predict position on a new set of times t_k.
2. Find the most compact representation to discover the physics of the system.

While (1) can be achieved via the standard approach to time series, we will rely on variational autoencoders we discussed to tackle point (2). We will force the NN to minimize the number of neurons in the latent representation, and the expectation is to get the physics of the system through the inspection of these neurons after the training and after checking that the predictions are good enough.

The general case is that we don't know anything about the system we want to study, so our choice will be to start a minimal number of latent neurons to find the best representation. Then we will make the predictions, and in case the accuracy will be low on the test data, we will increase the number of nodes in the latent representations. Otherwise, we will look at the weights stored in the latent representation as a good model of our physics system. Those values will represent the physical values that fully describe the damped pendulum.

Figure 6.12 shows the general architecture of our VAE (b) to study the damped pendulum and compare it to the general approach we follow in physics model a system (a): we start from observations, build a mathematical representation, and use this representation to predict output in different cases.

In the case of autoencoders (Fig. 6.12b), the experimental observations are encoded in a compressed representation. The process of decoding is to predict positions of the pendulum (or a generic physical system) at specific times based on the learned representation. As we said, the representation is called latent representation, and our main focus is on the representation in order to understand how much this representation is able to learn and reproduce the physics of the systems. Just predictions on pendulum positions could have been obtained with a variety of ML models, but what we are doing here is to constraint the number of nodes in the latent representation to get the physics. As an example, we may obtain 99.99% accurate predictions with a deep multi-layer neural network, but such an architecture would not allow the discovery of physics.

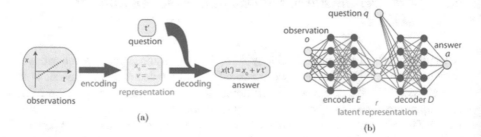

Fig. 6.12 Learning representation with human learning (**a**) vs SciNet neural network (**b**) (Iten et al. 2020)

We can have a brief look at the code, focusing on the most important parts. The code is taken and adapted from the work we already cited (Iten et al. 2020) and the implementation performed by Dietrich (2020).

For our purposes, we use the trained model provided by the authors and then comment on the results.

```python
import torch
import numpy as np
from models import SciNet
from utils import pendulum
from matplotlib import pyplot as plt
from mpl_toolkits.mplot3d import Axes3D

# Load trained model
scinet = SciNet(50,1,3,64)
scinet.load_state_dict(torch.load("trained_models/scinet1.dat"))

# Set pendulum parameters
tmax = 10
A0 = 1
delta0 = 0
m = 1
N_SAMPLE = 50
for ik, k in enumerate(np.linspace(5,10,size)):
    for ib, b in enumerate(np.linspace(0.5,1,size)):

        tprime = np.random.uniform(0,tmax)
        question = tprime
        answer = pendulum(tprime,A0,delta0,k,b,m)
        if answer == None:
            continue
        x = np.linspace(0,tmax,50)
        t_arr = np.linspace(0,tmax,N_SAMPLE)
        x = pendulum(t_arr,A0,delta0,k,b,m)
        combined_inputs = np.append(x, question)
        results = scinet.forward(torch.Tensor([combined_inputs]))

        latent_layer = scinet.mu.detach().numpy()[0]


        neuron_activation[0][ik,ib] = latent_layer[0]
        neuron_activation[1][ik,ib] = latent_layer[1]
        neuron_activation[2][ik,ib] = latent_layer[2]
```

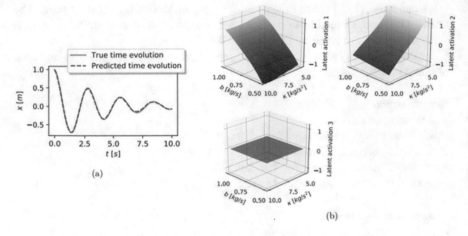

Fig. 6.13 Damped pendulum (**a**) Real evolution compared to the predicted trajectory by SciNet (**b**) Activation plot, representation learned by SciNet using the three latent neurons (Iten et al. 2020)

SciNet is the name of our NN, and in the line, in red, we see how to collect the results through torch open source library.

For us, as we said, it is important to investigate the latent layer (assuming that the predictions are good, as we will see) to look at the relevant physical variables.

Figure 6.13 shows the true time evolution vs the predicted time evolution of the damped pendulum and the representation learned by SciNet. The activation plots (b) clearly show as SciNet uses two of the three neurons in the latent representation to store the spring constant k and the damping factor b. The third neuron is not used as further confirmation of the fact that all the physics of the damped pendulum is "condensed" in the two meaningful variables of the latent representation.

Box below (Fig. 6.14) that is directly extracted from Iten et al. (2020) is very useful to summarize the problem and the findings:

This is an "easy" case that shows the general ideas about how to get knowledge discovery with ML and XAI. As properly commented by Iten et al. (2018), this is not a full solution of how to get an explanation about the latent variables. In our terminology, we are trying to get to rung three of the ladder, but we are not yet there. In this specific case, we got knowledge discovery but comparing the latent representation to the well-known model of the damped pendulum we have from physics. More generally, we may want to get this knowledge discovery through a learned representation but without any guidance from physics. We will see in Sect. 6.4 the most promising directions to do science in this way. Despite these limitations, we think it is pretty impressive to see how we can change the perspective and look at the ML models not just to have predictions but to gain insight and understanding of the system we are studying.

Problem: Predict the position of a one-dimensional damped pendulum at different times.

Physical model: Equation of motion: $m\ddot{x} = -\kappa x - b\dot{x}$.

Solution: $x(t) = A_0 e^{-\frac{b}{2m}t}\cos(\omega t + \delta_0)$, with $\omega = \sqrt{\frac{\kappa}{m}}\sqrt{1 - \frac{b^2}{4m\kappa}}$.

Observation: Time series of positions: $o = [x(t_i)]_{i \in \{1,...,50\}} \in \mathbb{R}^{50}$, with equally spaced t_i. Mass $m = 1$kg, amplitude $A_0 = 1$m and phase $\delta_0 = 0$ are fixed; spring constant $\kappa \in [5, 10]$ kg/s^2 and damping factor $b \in [0.5, 1]$ kg/s are varied between training samples.

Question: Prediction times: $q = t_{\text{pred}} \in \mathbb{R}$.

Correct answer: Position at time t_{pred}: $a_{\text{cor}} = x(t_{\text{pred}}) \in \mathbb{R}$.

Implementation: Network depicted in Figure with 3 latent neurons.

Key findings:

- *SciNet* predicts the positions $x(t_{\text{pred}})$ with a root mean square error below 2% (with respect to the amplitude $A_0 = 1$m)
- *SciNet* stores κ and b in two of the latent neurons, and does not store any information in the third latent neuron

Fig. 6.14 Summary of the findings. (Adapted from Iten et al. (2020))

6.3.3 Climbing the Ladder of Causation

We used the real case scenario of damped pendulum to touch with hands what it really means to climb the ladder of causation and go from the level of pure association to the rung of using ML in the domain of knowledge discovery. In order to further reinforce this approach, we look at it from a different angle exposed by Karim et al. (2018).

The three levels proposed by these authors exactly match the three rungs of the ladder of causation but from a mathematical perspective (Fig. 6.15).

Looking at Fig. 6.15, we see the flow from data to scientific knowledge:

1. *Statistical modeling*: this is the level of correlation, the ML model just learn these correlations from the data, then XAI may help on answering questions about the most important features and the sensitivity of the output to a specific change in a feature. As we see, from a mathematical point of view, it corresponds to $p(y|x)$.
2. *Graphical causal modeling*: this level receives as input the subset of features that have been already distilled and filtered from the statistical modeling as the most important ones. We enter into the domain of "interventions" as mathematically expressed by the $do(x)$ as in the rung two of the ladder of causality.
3. *Structural equation modeling* is the last level that deals with counterfactuals. It received filtered causes emerged from the causal modeling and put them into the form of structural equation modeling to produce knowledge. In our example this meant to go from time series up to the physical model of damped pendulum.

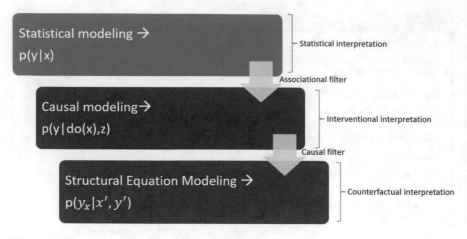

Fig. 6.15 Three-tier explainability (Karim et al. 2018)

The main idea behind this approach is to evolve XAI from a pure engineering approach to the possibility of doing science with ML. In the next section, we will see some options that are emerging to effectively use ML in scientific discovery.

6.4   Science in the Age of ML and XAI

The goal of this section is to discuss about the general ideas and options available to do science with ML and XAI. As we saw, touching with hands the case of damped pendulum, climbing the ladder of causality up to rung three to do real knowledge discovery is not easy. At the same time, we put the foundations to know what we need and how autoencoders may help in this direction. Also, in cases in which we cannot devolve to ML the task of discovering the physics of a system, there are other possibilities that may be worth to explore.

In general terms there are two main cases when we face the problem of doing knowledge discovery of physical world:

We have a lot of data, but we don't have any existing theoretical model of the system we want to study.

We have data and also a good mathematical model of the system.

In the first case, ML is already widely adopted to do predictions, which is the usual path: give me raw data, don't worry about any theoretical understanding of the numbers, use the data to train a ML model, and the predictions will come as a magic. But we can do more than this.

ML is already good in finding patterns, and this could be very useful in the territory of unknown where ML could point out meaningful directions to investigate in cases where we don't have any scientific theory available. But also, using the techniques of constraining the latent representation of VAEs' ML might be used to find hints of new physics (e.g., data from LHC in particle physics) (Guest et al. 2018).

Another option is also very interesting: use VAEs to do anomaly detections. Using again particle physics as example, it is already very common to do analysis of the huge number of data coming from accelerators as LHC in order to track the events of particle collisions.

As we learned autoencoder is basically a neural network that maps an input back to itself (with some degree of approximation) using a compressed latent representation. The same principle may be used the other way around: all the cases in which autoencoders fails to reconstruct an event can be taken as an anomaly – a new class of event that we should study to look for new physics. We are making the things too easy (indeed there is a lot of work to do as explained by Farina et al. (2020)), but the objective here is just to give you the feeling of the different approaches to do science with ML.

Let's now think about case 2: we have a lot of data but also a mathematical model; does ML help in this case? The answer is yes through a hybrid approach. Think about very complex physical systems in which doing "real-time" predictions with the mathematical model is not feasible because it would take too much time or computational resources. The approach may be to use the mathematical model to generate enough data to train a ML model and then use the ML model to do real-time predictions instead of the mathematical model for better performance.

Following this path, the reader may ask the ultimate question: is there any general and profound reason why the NN works so well in doing prediction for most of the problems after suitable training? Said in other terms, in principle, neural networks are able to approximate any function, but you may expect that it would be very hard and computationally complex to learn systems in which functions may have arbitrary polynomial orders, while the reality is that usually low polynomial orders (or simple functions) are enough to have good results on the majority of problems. This is an open stream of research with deep implications that are beyond Machine Learning and touch fundamental physics. The scope of this book is not to deep dive these arguments, but we think it is really beautiful to follow at least the basic ideas as explained by Lin et al. (2017):

1. "For reasons that are still not fully understood, our universe can be accurately described by polynomial Hamiltonians of low degree." The laws of physics have orders ranking from 2 to 4, and this reduces the functions that need to be approximated by NN.
2. Modern theoretical physics is guided by mathematical symmetries that constraint the underlying mathematical model. This in turn helps the ML during the training: think about image recognition, images of dogs or cats are symmetrical, and this strongly simplifies the learning phase.
3. The hierarchical structure of any physical system may play a role also on NN side. Elementary particles build atoms that build molecules up to planets and galaxies. The layers of NN could mimic this "inner" order in their sequential and causal approximation of the results adding complexity or details in each layer.

We want to stress again that these ideas are mainly speculative, but we liked them a lot, and we think that they deserve some reflection. Because of this we thought to share with you readers even if we know that we cannot touch them with hands as we did with the rest of techniques and tools we have discussed so far.

6.5   Summary

- Adapt scientific method to the current times with the huge availability of data.
- Understand the ladder of causation to formulate the right questions, and use the right XAI tools to answer.
- Use autoencoders to build representations of physical systems.
- Discover the physics of damped pendulum with variational autoencoders.
- Learn the basics to do science with hybrid approach using ML.
- Use autoencoders to detect anomalies.
- Get insight into the power of NN to approximate different physical systems.

References

Dietrich, F. (2020). *Implementation of SciNet*. Available at https://github.com/fd17/SciNet_PyTorch.

Farina, M., Nakai, Y., & Shih, D. (2020). Searching for new physics with deep autoencoders. *Physical Review D, 101*(2020), 075021.

Gilpin, L. H., Bau, D., Yuan, Z. B., Bajwa, A., Specter, M., & Kagal, L. (2018). Explaining explanations: An overview of interpretability of machine learning. *arXiv:1806.00069*, arxiv.org.

Guest, D., Cranmer, K., & Whiteson, D. (2018). *Deep learning and its application to LHC physics*. https://doi.org/10.1146/annurev-nucl-101917-021019.

Iten, R., Metger, T., Wilming, H., Del Rio, L., & Renner, R. (2020). Discovering physical concepts with neural networks. *Physical Review Letters, 124*, 010508.

Karim, A., Mishra, A., Newton, M. A., & Sattar, A. (2018). Machine learning interpretability: A science rather than a tool. *arXiv preprint arXiv:1807.06722*, arxiv.org.

Lin, H. W., Tegmark, M., & Rolnick, D. (2017). Why does deep and cheap learning work so well? *Journal of Statistical Physics, 168*, 1223–1247.

Pearl, J., & Makenzie, D. (2019). *The book of why* (eBook ed.). UK:Penguin.

Chapter 7
Adversarial Machine Learning and Explainability

"If you torture the data long enough, it will confess to anything."

—*Ronald Coase*

This chapter covers:
What is Adversarial Machine Learning?
Doing XAI with adversarial examples.
Preventing adversarial attacks with XAI

Let's get into the main topic of this chapter with an impressive example, looking at Fig. 7.1:

Do you see any difference between these two pandas? I bet the answer is no; we don't have any doubt on saying that both of them represent a panda. But as shown by Goodfellow et al. (2014), the first one has been classified as a panda by a NN with 55.7% confidence, while the second has been classified by the same NN as a gibbon with 99.3% confidence. What is happening here? The first thoughts are about some mistakes in designing or training the NN, but the point that will emerge from this chapter is that this mistake in classification is due to an adversarial attack. As we will learn, it is pretty easy to fool the neural networks with a variety of adversarial attacks that, with some intangible, for the human eye in case of image classification task, change in the input, can break the classifier.

Even if we may find this topic interesting enough to be deeply dived, the obvious question that follows is to understand how Adversarial Machine Learning is related to XAI, which is the topic of this book. In the next sections, we will see that the relation between XAI and Adversarial Machine Learning is twofold: on one side, XAI can be used to make the ML models more robust and prevent adversarial attacks, while, on the other side, adversarial examples can be considered as a method to produce local explanations (among the other XAI techniques we already discussed).

L. Gianfagna, A. Di Cecco, *Explainable AI with Python*,
https://doi.org/10.1007/978-3-030-68640-6_7

Fig. 7.1 Comparing two pictures of a panda. (Goodfellow et al. 2014)

But before exploring the relationship between AE and XAI, we start with a crash course on Adversarial Machine Learning to set the foundations.

7.1   Adversarial Examples (AEs): Crash Course

The story of AE begins in 2013 with the seminal work of Szegedy et al. (2013) in which the authors focus on two properties of neural networks: the first one regards how neural networks work, while the second property investigates the stability of the NN against small perturbations of the input. The paper was not meant to deep dive adversarial examples, but the results related to the study of the second property set the foundations for Adversarial Machine Learning.

At that time, there was a lot of enthusiasm around NN's visual and speech recognition performances that were achieving results similar to human counterparts. The belief was that accuracy was naturally coupled with the robustness of the neural network. The shocking result of Szegedy et al. (2013) was instead that an imperceptible non-random perturbations to the input image might change the network's classification of the image almost arbitrarily, and this is achieved by optimizing the input to maximize the prediction error. Quoting the authors: "We term the so perturbed examples as *adversarial examples (AE)*."

The results of the paper are summarized in Fig. 7.2 that is extracted from the paper itself:

Figure 7.2 shows in the left columns the original images distorted by perturbation, shown in the middle column, to produce the adversarial example in the right column.

All the images in the right column are classified as an ostrich, although the perturbation is not visible for the human eye.

These adversarial examples have been generated for AlexNet – the name of the convolutional neural network designed by Alex Krizhevsky. This CNN won the

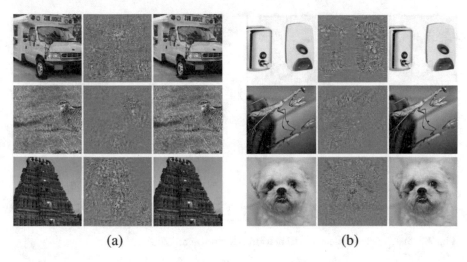

(a)   (b)

Fig. 7.2 Left columns are original images. In the middle we have the perturbation; on the right there are the hacked images incorrectly classified. (Szegedy et al. 2013)

ImageNet Large Scale Visual Recognition Challenge on 30 September 2012 with a top-5 error of 15.3%, more than 10.8 percentage points lower than that of the runner up. The paper shows how to fool other state-of-the-art networks besides AlexNet.

Moreover, quite unexpectedly, at the time, a huge fraction of the generated AE is also misclassified by networks trained from scratch with different hyper-parameters. This behavior is the first indication of the possible "transportability" of the attacks discussed in the following.

What is the general method to produce these adversarial examples? Let's try to highlight the main steps to generate them. Suppose to have an image X classified by the NN as class A using the function $f(x) \rightarrow A$. We want to find the smallest perturbation d so that x is classified as B (that is different from class A) by the NN; formally we search for d so that:

$$\min_d d = \|d\|_2 \tag{7.1}$$

$$f(x+d) = B \text{ with} (x+d) \text{ being a valid image}$$

This is an optimization problem that is not easy to solve. The first approach was to use the so-called L-BFGS algorithm (i.e., a variation of the original BFGS algorithm named with the initials of the authors) that somehow limits the cases to which these attacks can be applied.

L-BFGS entered the literature as the first AE showing how the NNs are less robust than the common beliefs. The "brittle" term emerged as opposed to robust and has been applied to NNs fooled by these attacks. Do you remember Fig. 1.3 of Chap. 1 (repeated in Fig. 7.3)?

Fig. 7.3 State-of-the-art deep neural network. (Nguyen et al. 2015)

We used this image to introduce XAI's need to understand the features on which the NNs are relying for the classification and avoid cases like the ones shown in the figure. Looking at the same case, we see how these AEs may fool the NN, and XAI could be used to avoid such a state of things (as anticipated, we will discuss the twofold relation between XAI and AE).

After the original paper started the AE, the research progressed to understand if AEs were rare cases or could be easily generated. Moreover, it was essential to know if the knowledge of the internals of NN was strictly required to create an AE or not. L-BFGS was used to craft AE, but it is a general optimization algorithm, and it doesn't shed light on the phenomenon. All these points were prerequisites to answer the most important question about preventing NNs from being victims of AEs.

The answers didn't take too long to come. In 2014, Goodfellow et al. achieved two fundamental results:

1. A technique to directly generate AEs.
2. The understanding of root cause of the vulnerability of general NNs to adversarial examples.

We are taking a historical perspective in this first part of our crash course into AE as we want to place in the proper context the main results that made AE jump out of purely academic interest as a potential threat to the wide adoption of Machine Learning.

The first reaction of the same authors of "intriguing properties of neural networks" (Szegedy et al. 2013) was a kind of uncertainty about the scope and extent of AE as outlined in their conclusion in which they questioned how the NNs could be fooled by these attacks, quoting their own words:

The existence of the adversarial negatives appears to be in contradiction with the network's ability to achieve high generalization performance. Indeed, if the network can generalize well, how can it be confused by these adversarial negatives, which are indistinguishable from the regular examples? Possible explanation is that the set of adversarial negatives is of extremely low probability, and thus is never (or rarely) observed in the test set, yet it is dense (much like the rational numbers), and so it is found near every virtually every test case. However, we don't have a deep understanding of how often adversarial negatives appears, and thus this issue should be addressed in a future research. (Szegedy et al. 2013).

As shown by Goodfellow et al. (2014) and following research, AE appears quite often and easily.

The best way to look at the two points above is to understand why the AE can be generalized, and based on this we will give some arguments about how to generate them quickly. The fact that AEs are generalized so easily and that different architectures of NNs may be vulnerable to the same AEs was a kind of surprise for the ML researchers. The initial explanations about this point were along the direction of extreme nonlinearity of the deep NNs combined with some lack of required regularization and overfitting in some cases. Ironically enough, Goodfellow et al. showed exactly the contrary: *the vulnerability of NNs to AE is mostly due to a combined effect of extended linearity of NNs and high-dimensional input.*

Assuming to have a linear model (I know your question about why we should assume the inner linearity of deep NNs, and we will get back to this), every feature is defined with an intrinsic precision. The classification is expected to give the same classification to two different inputs x and x' if

$$x' = x + \varepsilon \qquad (7.2)$$

if every element of the perturbation ε is less than the precision of the features.

If we consider in general the dot product of the weight vector (w) and the adversarial example x', we have:

$$w^T x' = w^T x + w^T \varepsilon \qquad (7.3)$$

The perturbation shifts the activation by $w^T \varepsilon$. But w is in general highly dimensional; assuming to have n dimension and m as the average value of an element of the weight vector, the shift is about $m * n * \varepsilon$. So we understand how keeping ε small can, in any case, have a large change in the activation because the overall growth grows linearly with the dimension of the problem that is n, which means obtaining a large difference in the output with small changes in the input. Let's look at the linearity from a visual angle to touch with hands why linearity is responsible for the vulnerability against AE.

Linear models extrapolate data without any flattening in regions where we don't have so much data. Each feature keeps the same partial slope across space, without any dependence or consideration of the other features. Said in other terms, if you are able to push a bit the input in the right direction to cross the decision boundary, you can easily reach the space to get a different classification.

Figure 7.4 shows this behavior in which choosing the right vector in the orthogonal direction to the decision boundary quickly takes the model out of the original classification. Based on these arguments, we can introduce a more efficient

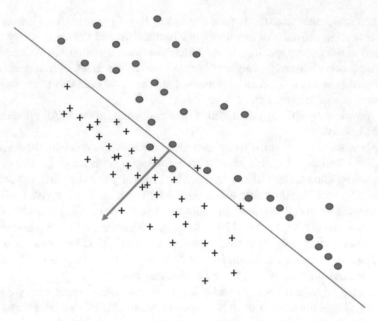

Fig. 7.4 How to cross the boundary decision of a classification

algorithm to generate AE: the so-called FGSM (fast gradient sign method). Back to the starting optimization problem:

$$\min_d d = \|d\|_2 \qquad (7.4)$$

$$f(x+d) = B \text{ with}(x+d) \text{being a valid image}$$

L-BFGS is the algorithm that is used to solve it, and in general it is computationally expensive.

FGSM approach is different, and attacks can be generated more easily. It is interesting to understand the foundations of FGSM because it helps a lot to get the basics of AE (Fig. 7.5).

Let's consider the usual gradient descent in one dimension in which we want to find the minimum of the loss function.

The optimization is performed to find the best weights for the NN to get the minimum of the loss function. Just considering one specific data point, it looks like this:

$$L(x,Y,\theta) = \left(f_\theta(x) - y\right)^2 \qquad (7.5)$$

In this specific case, we are searching for the optimal θ values to get the minimum. At each iteration of the gradient descent algorithm, θ is updated toward the minimum:

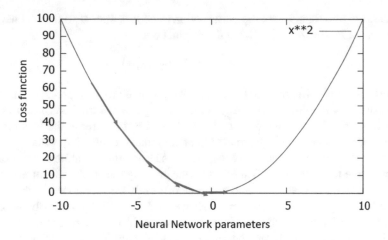

Fig. 7.5 Loss function in one dimension

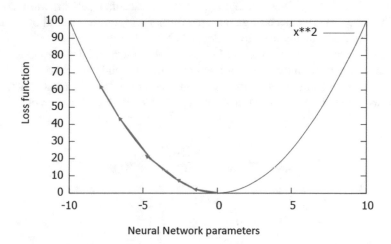

Fig. 7.6 Gradient descent and FGSM

$$\theta' = \theta - \alpha \nabla_\theta L(x,Y,\theta) \tag{7.6}$$

The idea of FGSM is to use the same approach of gradient descent but work on the data points instead of the parameters. Our minimum problem is to find the values for x to increase the loss for that specific example:

$$x' = x + \alpha \nabla_x L(x,Y,\theta) \tag{7.7}$$

We keep fixed the parameters of the model and differentiate with respect to the specific input x in the other direction if compared to the gradient descent algorithm (Fig. 7.6).

FGSM is just this: we fix the perturbation to be less than ∈ so that the new data point cannot be distinguished from the original one:

$$x' = x + \varepsilon \, sign \nabla_x L(x, Y, \theta) \tag{7.8}$$

There is no specific optimization here. We just need to set ε and apply the same perturbation to the data points considering the sign of the gradient used to understand if we need a positive or negative perturbation to increase the loss function. But, as we explained, it is constrained to be less than ε in absolute value.

With this method, we can quickly generate AE, and the literature shows how it is pretty simple to reach up to 90% errors on the primary image classification datasets.

We still need to dive deep into a couple of points we mentioned in this section before touching on how to generate AE. We said that AE's power is based on the intrinsic linearity of the NN that is pretty shocking for every ML student. We say surprising because the basic theory about Deep Neural Networks is that their impressive performance is mostly due to their deep structure of nonlinear activation functions. That's why they can learn functions that cannot be learned by shallow and linear NN. But we need to go one step further and revise the kind of nonlinear functions that are usually used to model deep NN.

One of these functions is the ReLU that is well known and depicted in Fig. 7.7:

ReLU is linear for a large part of its domain ($x > 0$), and this makes ReLU enormously different from the other two functions that are commonly used as activation functions in DNN: the logistic function and the bipolar (tanh) function of Fig. 7.8.

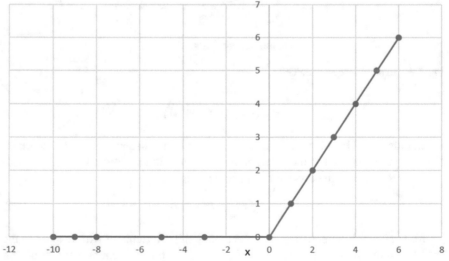

Fig. 7.7 ReLU function

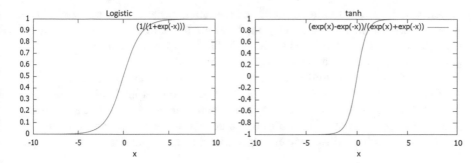

Fig. 7.8 Logistic and tanh activation functions

Both logistic and tanh (bipolar) activation functions exhibit a saturation behavior that is not in ReLU. This asymptotic behavior makes these functions more robust against AEs but more difficult to train. The saturation makes the gradients very close to zero, and so it is more difficult to train them outside of the region where they are almost linear. But this same reason makes them more robust against AEs because the nonlinear capping depresses the overconfidence that we mentioned of also extrapolating in regions where there are not so many data points.

The fact that the training and computations (in particular for the ReLU that just needs a sign check for computation) are easier in the linear regions explains the general vulnerability of NN to AEs. Also, Deep Neural Networks are trained for a large part in a linear regime that makes them vulnerable to AE. Quoting Goodfellow et al. (2014): "More nonlinear models such as sigmoid networks are carefully tuned to spend most of their time in the non-saturating, more linear regime for the same reason. This linear behavior suggests that cheap, analytical perturbations of a linear model should also damage neural networks...[There is] a fundamental tension between designing models that are easy to train due to their linearity and designing models that use nonlinear effects to resist adversarial perturbation. In the long run, it may be possible to escape this trade-off by designing more powerful optimization methods that can successfully train more nonlinear models." So far, we understood why NN are so vulnerable to AE and how to easily generate them. But before touching with hands AE and then exploring the link between AE and XAI, we still miss a fundamental piece of the puzzle: what we did so far assumes that we have access to the neural network models so the AE we talked about would not apply to NN you cannot access. It seems that you cannot just perform black-box AE on existing NNs that are exposed on the cloud and in general without knowing the internals (as we saw the gradient values at least) of NN itself. But the situation quickly worsened with the further seminal work of Goodfellow et al.: AEs are not only universal in the sense that can be used to attack whatever type of NN, but they can also be easily transferred from one NN to another; it was the birth of "black-box" AE.

The idea in general terms is easy to get: the only assumption is to have access to the NN we want to attack but only to look at the labels (classifications) provided to specific inputs. Then a local model is trained to replace the target DNN. The training is performed using a synthetic input and the labels generated by the target DNN when exposed to this input. Having the local model can then be used to craft AE using the techniques we learned so far in the "local" space of inputs in which it is a good approximation of the target DNN to be attacked.

This work from Goodfellow et al. (2014) opened the door to further evolutions starting from universal perturbations (Moosavi-Dezfooli et al. 2017) to the recent one-pixel attacks that showed how to fool a neural network by just changing one pixel in the input image. We won't go into details into these technologies but just sketch the ideas behind the universal perturbations to emphasize the main results (Fig. 7.9).

1. Given a distribution of images d and a classification function, it is possible to find a perturbation that fools the classifier on almost all the images sampled from d. Such perturbation is called universal as representing a fixed image-agnostic

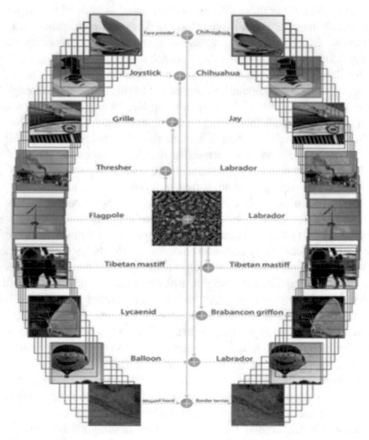

Fig. 7.9 Left images are the original images with proper labels; central image is the universal perturbation; on the right there are the misclassified related images because of the universal perturbation. (Moosavi-Dezfooli et al. 2017)

Table 7.1 Generalizability of the perturbation across different architectures; the numbers are the fooling rates, and the max values are reached as expected along the diagonal (perturbation computed for an architecture and applied to the same architecture). But the generalizability is pretty huge also across the off-diagonal cells (Moosavi-Dezfooli et al. 2017)

	VGG-F	CaffeNet	GoogLeNet	VGG-16	VGG-19	ResNet-152
VGG-F	93.70%	71.80%	48.40%	42.10%	42.10%	47.40%
CaffeNet	74.00%	93.30%	47.70%	39.90%	39.90%	48.00%
GoogLeNet	46.20%	43.80%	78.90%	39.20%	39.80%	45.50%
VGG-16	63.40%	55.80%	56.50%	78.30%	73.10%	63.40%
VGG-19	64.00%	57.20%	53.60%	73.50%	77.80%	58.00%
ResNet-152	46.30%	46.30%	50.50%	47.00%	45.50%	84.00%

perturbation that causes the change in classification for the images in the sample, keeping the perturbed images almost indistinguishable from the original ones.

2. The universality is twofold: the perturbation is not only universal across different data points but also across different architectures of NN. As shown in the paper (Moosavi-Dezfooli et al. (2017)), the perturbations generalize pretty well across the six architectures that have been tested (Table 7.1).

This second kind of universality indeed can be considered just as an experimental confirmation of our theoretical discussion: we don't need to shape an AE for each specific DNN architecture given the fact that whatever is the DNN architecture, the DNN spends most of their training in the linear regime that is at the root of their same vulnerability to AE. Let's go into a practical example of how to craft AE before exploring AE as a XAI technique and how to defend from AE using XAI.

7.1.1   Hands-On Adversarial Examples

As promised, we will use this section to show how to craft an easy AE.

Before coding, let's get familiar with the math we talked about with a toy model (Karpathy 2015). Suppose to have a basic logistic classifier that gives as output 0 or 1 depending on two possible classes.

$$P\left(Y=1 \mid x; w, b\right) = \sigma\left(w^{t} x + b\right) \tag{7.9}$$

Receiving x as input, the classifier assigns x to class 1 if $P > 50\%$. σ is the standard sigmoid function that maps the combination of weights and inputs (w and x, $b = 0$) between 0 and 1.

Suppose we have the input and the weight vector w below:

```
x = [2, -1, 3, -2, 2, 2, 1, -4, 5, 1] // input
w = [-1, 1, 1, -1, 1, -1, 1, 1, -1, 1] // weight vector
```

Doing the dot product, we get -3, which means that the probability to have this input classified as class 1 is $P = 0.0474$ which is low. This input would be classified as class 0 with a probability of about 95% which is pretty strong.

We now use FGSM to tweak this input; remember that the idea behind FGSM is to have small changes in the input to make the overall image (or whatever else input) indistinguishable from the original one and at the same time change the resulting classification.

And to achieve this goal, FGSM recommends to set a small eps and perturb the input along with the same sign of the weight (positive if positive and negative if negative):

$$x' = x + \varepsilon \operatorname{sign} \nabla_x L(x, y, \theta) \qquad (7.10)$$

In this toy example, set $\varepsilon = 0.5$ and change the input (named adx) accordingly:

```
adx = [1.5, -1.5, 3.5, -2.5, 2.5, 1.5, 1.5, -3.5, 4.5, 1.5]
```

If we do the dot product again with this, it gives 2 (instead of -3) this time, and computing the overall probability we have that the image is classified as class 1 with $P = 0.88$ (instead of 0.0474) which means that it will be assigned to class 0 (instead of class 1) with 88% probability.

What is the takeaway of this dummy example? That is, just changing the input of eps = 0.5, we got an overall effect that strongly influenced the overall probability. This is due, as we saw in the theoretical discussion, to the number of dimensions and the dot product that amplifies the effect of a small perturbation. Consider also the fact that, usually, we have thousands of dimensions instead of just ten we used to touch with hands what is going on, and given what we saw in this example, a very small eps may cause even bigger changes in the classification keeping the input globally indistinguishable from the original one.

In real-life scenarios, there are a lot of libraries to quickly generate AE. We will use Foolbox – Python library – that is used to create adversarial attacks against the majority of Machine Learning models like a Deep Neural Network. This library works natively with models built in PyTorch, TensorFlow, and JAX; for our purpose, we will discuss and comment on what is provided as an example in the documentation using PyTorch (Foolbox 2017). We use the pre-trained model ResNet18 that is a convolutional neural network that has been trained on images from the ImageNet database. It is used to classify images into more than 1000 object categories spanning from animals to pencils.

In the following, we provide the main snippets of code to understand the flow. We start with the imports

```
import foolbox
import torch
import torchvision.models as models
import numpy as np
```

to get what is needed in terms of Foolbox and PyTorch.
The next step is to instantiate the model.

```
resnet18 = models.resnet18(pretrained=True).eval()
if torch.cuda.is_available():
 resnet18 = resnet18.cuda()
mean = np.array([0.485, 0.456, 0.406]).reshape((3, 1, 1))
std = np.array([0.229, 0.224, 0.225]).reshape((3, 1, 1))
fmodel = foolbox.models.PyTorchModel(
resnet18,bounds=(0,1),num_classes=1000,preprocessing=(mean,std))
```

and get the image we want to attack

```
# get source image and label
image,    label    =    foolbox.utils.imagenet_example(data_format='
channels_first')
image = image / 255. # because our model expects values in [0, 1]

print('label', label)
print('predicted class', np.argmax(fmodel.predictions(image)))
```

The print statements are to check what is loaded against what is predicted, that
is, class 282 corresponding to a tiger cat from the ImageNet database.
The last step is just to create the AE with just two lines of code using the Foolbox
library:

```
# apply attack on source image
attack = foolbox.attacks.FGSM(fmodel)
adversarial = attack(image, label)
```

Adversarial here is the image that has been manipulated with FGSM attack. If we
look into the code to generate the attack, we find what is expected in terms of our
previous discussions which is something like this:

```
perturbedImg = img + gradient_sign * np.sign(a.gradient())
```

where a.gradient() automatically evaluates the generic $\nabla_x L(x, Y, \theta)$ term of FGSM.
And checking the results with the print below.

```
p  r  i  n  t  (  '  a  d  v  e  r  s  a  r  i  a  l
class', np.argmax(fmodel.forward_one(adversarial)))
```

the output is 281 which means that our convolutional neural network is now
wrongly classifying a tiger cat (image 282) as a tabby cat (Fig. 7.10).

Tiger cat (id:282)   Perturbation   Tabby cat (id:281)

Fig. 7.10 Adversarial attack example: tiger cat misclassification

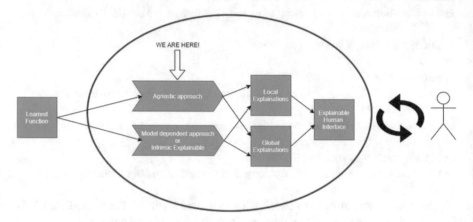

Fig. 7.11 XAI flow

7.2   Doing XAI with Adversarial Examples

As we said, the relation between XAI and AE is twofold: in this section, we show
how AE can help in doing XAI, while in the next section, we will see how XAI can
be used to make ML more robust against AE. Going into these details will allow
understanding what the root of this relation between XAI and AE is.

Back to our picture of the XAI flow (Fig. 7.11).

We saw how the methods to do XAI might produce local or global explanations.
There is a trending technique in XAI that we didn't explicitly mention so far, but
that might be included in the family of agnostic approach: the so-called example-
based explanations (Adadi and Berrada 2018). Example-based explanations mean
interpreting the model choosing the most representative instances in the dataset to
represent the model behavior (Molnar 2019). Like model-agnostic methods, we
don't need any access to the model internals, but differently from model-agnostic
methods, there is no attempt to summarize or to narrow down the most relevant
features. In this sense, example-based explanations are more meaningful for humanlike kind of explanations: as humans searching for explanations often means to
search for an example, a case that makes it simple to understand what's going on.

The path is that if two events are similar, we usually get to the conclusion that they will generate the same effect.

If a loan is refused to one of our friends, we try to compare our situation to his to understand what the criteria are for a loan to be granted or not. The example is more tangible than trying to understand and get explanations for the whole ML algorithm. There two main types of example-based explanations: the prototypes and counterfactual explanations. We will see how AE can be considered a specific case of counterfactual explanations.

Prototypes are what you might expect precisely from the meaning of the word: we search for instances looking from the most representative ones. The explanations are then generated by looking at the similarity of other data points to those chosen as prototypes. To avoid generalizations, a prototype is often coupled with a "criticism" which, on the contrary, is a data point that is not well represented by any prototype. Prototypes are usually identified through clustering algorithms like k-means.

We mentioned prototypes for completeness, even if we are more interested in counterfactual explanations and their relation with XAI. Counterfactual explanations mean to search for the minimal conditions for the minimal changes to apply to a specific input that would have caused a different decision for that input. In our example, assuming to have a loan that has been accepted, we would search for the smallest change in one or more features to have the loan rejected.

This is again an example-based explanation, but it is different from prototypes. Prototypes are data points that exist in the dataset, while a counterfactual example is a data point that is not present in the dataset, an event that didn't happen, something in which the ML model has not been trained or tested.

Following this path, we can now understand how this is related to adversarial examples that can be considered a specific type of counterfactual example-based explanation. Do you remember what an AE is? We searched for the minimal changes for a data point to fool a ML model. That is precisely the same kind of reasoning we followed to explain counterfactual explanations. Learning to fool a NN can be considered a XAI method in the sense that it is learning what to touch; how to change the features to change the prediction is to learn how the ML model works. As for agnostic methods, considering the ML as a black box, the creation of AE helps us to understand "why" and "how" a specific input received a specific classification and how this input would be sensitive to a small change in the most important features that produced that classification.

We said that differently from prototypes, counterfactual instances do not exist in the dataset. This should be a kind of deja vu for the reader because we deep dived counterfactual reasoning from a different angle in Chap. 6 already. It is worth to recall some important topics that we already covered there to have the full picture.

We talked about the ladder of causation that is part of Pearl's seminal work on causation (Pearl and Makenzie 2019).

We know that counterfactuals are at the top of the ladder, and to get there, we need to deal with imaging and retrospective climbing from interpretability to full explainability (remember that we saw in Chap. 6 that the law of physics can be considered a kind of counterfactual assertions).

Doing XAI with counterfactual example-based explanations is a human-friendly method of producing explanations because humans are naturally inclined to make sense of things answering questions like, "What if I had acted differently?". The alert about this is that the world in which something went differently does not exist, and so we need to get a full causal model to deal with such a state of things.

Having this background in mind and having identified AE as a specific type of counterfactual example-based explanation, we can look at how to generate counterfactual explanations with the methods we learned in XAI.

We saw in Chap. 4 the power of SHAP method to generate explanations for a single instance. We can adopt the same approach in this context and use SHAP to generate counterfactual explanations.

Remember that the Shapley value Ψ_{ij} for feature "j" and instance "i" is how much the specific feature contributed to the classification of instance "i" if compared to the average prediction of the dataset, so that Shapley values can be used to understand which factors contribute more or against a particular classification.

Following the work of Rathi (2019), we can use this algorithm to generate counterfactuals with SHAP. The idea is to use SHAP to answer P-contrast questions that are of the form, "Why [predicted-class] not [desired class]?". We want to deep dive for a specific data point; we calculate the Shapley values for every possible target class. The negative Shapley values contribute negatively to the target classification, while positive values do the vice versa. We can break the P-contrast question into two parts: Why P? Why not Q? We get Shapley values for P and Q classes and use those that work against the classification of the category selected to obtain counterfactual data points.

Given the data point, we estimate its Shapley values for each of the possible target classes. The negative Shapley values indicate the features that have negatively contributed to the specific class classification and vice versa.

The algorithm implements this flow: the starting point is to identify the desired class (Q), the predicted class (P), and the data point. Shapley values are calculated for each target class to produce counterfactual explanations. This approach has been tested on *Iris* dataset and Wine Quality dataset. As reported in the paper, in the *Iris* dataset, the answer to the basic question "Why 0, not 1" produced as an explanation that the petal width has strongly influenced the result 0 while the petal length has driven the counterfactual classification with label 1.

In this case, the explanation points out that to change the classification from 0 to 1, the target feature is the petal length. Generally speaking, we narrow down the features that work against the classification of the desired category; we can also get the counterfactual data that are related to contrastive explanations and give real examples of what needs to be changed to achieve a specific output. These data points represent the counterfactual answer to the contrastive query.

It is pretty different from the standard approach to generate AE as a counterfactual because the method we saw relied on a fixed $\in$ to provide a small but widespread perturbation of the features for the single data point. But the root of the problem is the same; we are challenging our methods to understand what makes a flower that specific kind of flower and discover how to change it.

In the following section, we will explore the other direction of the link and defend against AE using XAI (instead of doing XAI using AE) to close the loop.

7.3 Defending Against Adversarial Attacks with XAI

Keeping in mind what we learned about AE, the obvious question that emerges is how to defend against AE making the ML models more robust.

Our scope is to investigate the relation between AE and XAI, and we already saw how AE could be considered a specific type of example-based explanations like counterfactuals. There are several approaches for defenses against AE, and we will focus on the one that uses XAI itself to defend against AE. Just to mention the general approaches, there are four main types:

1. *Data augmentation*: this idea is to add AE as part of the training to make the model more robust. In this way, the model is trained against that specific type of AE, but the apparent limitation is that this method would require the knowledge of all the possible attacks to be exhaustive.
2. *Defensive distillation*: in ML literature, distillation is used as a general method of reducing the size of DNN architectures to decrease the request on computational resources. The high-level idea is to extract information from the original DNN and transfer it to a second DNN of reduced dimensionality. Defensive distillation is a variation of this method to increase the robustness of a DNN against AE. This is obtained with a two-phase procedure similar to distillation but aimed to enhance resilience against perturbations instead of compression of the DNN.
3. *"Detector" subnetwork*: in this approach (on detecting adversarial perturbations) the original DNN is not changed, but a small additional detector subnetwork is trained on a binary classification to distinguish original input from the ones containing AE.
4. *Adversarial training*: this is one of the most promising approaches. In this approach, all the AEs found are used to augment the train set. This procedure can be applied recursively, obtaining increasingly robust models to AE attacks.

Using XAI methods to defend from AE is an additional emerging approach that cannot be classified in these four main families of defenses.

The background to understand how XAI can be used is outlined by the work of Ilyas et al. (2019) in which AE is considered as an intrinsic property of the dataset itself. The authors introduce the concept of robust and non-robust features. Non-robust features are features that are highly predictive but are very sensitive to every change in the input. We can consider them as details that would not be generically used by a human being to perform a classification task.

On the contrary, robust features are again highly predictive but are not impacted by small input changes. In the case of a car, we may consider the wheels' presence as a robust feature just to give an example. For our purposes, the behavior of robust/non-robust features against small changes of input is fundamental for AE. AEs are crafted searching for non-robust features so that a small change in the input produces a considerable change in the values of these highly predictive features. A direct attack to a robust feature would not be feasible because it would require a more significant change to the input easily discovered by an observer.

Given the above, Fidel et al. (2020) showed how to use SHAP to leverage the difference between robust and non-robust features and defend against AE. While in the previous section we used SHAP to generate example-based explanations (kind of AE), in this case, we use SHAP from the opposite direction to make the ML model more resistant against AE.

Considering that XAI's main objective is to interpret ML models providing the relative importance of the features in determining the output, the hypothesis is that we may apply XAI to discriminate AE from original inputs. The idea is that the classification of a normal input should rely more on robust features if compared to the classification of AE that is likely to rely on non-robust features attacked to change the output classification.

We use SHAP to have a ranking of the relative importance of features to identify AE leveraging the different SHAP signatures.

Figure 7.12 adapted from the work of Fidel et al. (2020) clearly shows the proposed solution. On the left and right parts of the figures are normal examples of cats and cars. In the middle there is an original input for a cat that is hacked with a PGD L2 attack to create an AE that has as target class a cat.

Up to now, there is nothing new, in the sense that, as expected, the original example of a car and the adversarial example are indistinguishable for a human observer.

But looking below the figures, we can see the SHAP signature is added to each figure so that the SHAP value of each neuron i for target class j is provided in

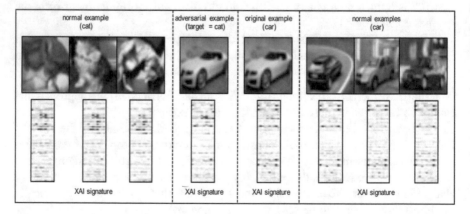

Fig. 7.12 Different images with different SHAP signatures. Normal and hacked examples are compared based on their SHAP signatures. (Fidel et al. 2020)

different colors. The red pixels are the positive contributions toward the target class; the blue ones are the negative contributions with an intensity that depends on the absolute value of the contribution itself.

In particular, the transparent pixels don't contribute to the classification that is what we see in the dedicated section of Chap. 4 for SHAP. This time we can look at these signatures from a different angle. A visual review of the figure is enough to get the main flow of the method.

Each image has a similar SHAP signature if compared with images of the same type. All the cats on the left and the cars on the right have a matching red pixel pattern (the positive ones).

The standard cars have five evident and strong rows, three rows on the top, one in the middle, and two on the bottom of the SHAP diagram. The cats on the left are in the same situation; the SHAP signatures are different from the cars, but they share the same red pixel pattern in the middle. The exciting part comes if we look at the SHAP signatures of the adversarial example. Out of the original five-strong rows of the original car image, the car AE kept only two rows, while we don't see any correspondence with the other 3 SHAP rows.

The cited paper shows how the three rows that disappeared in the AE are the non-robust features: AE attacked the non-robust features that are highly predictive but very sensible to small input changes. And the AE also shows the middle pattern of red pixels similar to the cat image. As for the car's original image, the AE transferred only the cat's non-robust features to avoid the robust ones.

The classifier produces the output for the AE based on a mixture of features coming from car and cat images, and the method proposed by the paper to recognize the AE is to rely on the SHAP signatures. We just presented the idea of Fidel et al. and the basic concepts. The paper further shows on real data how the method works in a real case scenario.

For our purpose, it is important to get a deep connection between XAI and feature robustness. We used SHAP to generate AE and the same SHAP method to detect AE narrowing down the non-robust features as SHAP signatures of the AE themselves.

Before closing this chapter, we would further emphasize the theoretical connection between AE and XAI. As we saw, most agnostic methods rely on input gradients to select the most important features for a model and produce explanations. And at the same time, we saw how AEs exploit these gradients to understand where the NNs are more vulnerable to small changes in the input. Adversarial gradients are directions where small perturbations generate big output variations, as we understood from our linear regimes discussion. To defend against such perturbation, the idea is to reduce the variation of the outputs around adversarial gradients to smooth the function learned by the NN to generalize better outside the domain of training. But smoothing these gradients means also making them more interpretable in explaining model predictions closing the loop between XAI and AE: robustness of the model against AE helps XAI, and XAI helps *defending against AE*.

This chapter is very rich in content and essential ideas that are not easy to digest. There is a lot of theory. Despite our attempts to make this theory backed up by real case scenarios and straightforward examples, we are aware that we are moving away toward a more "research"-like material. The most important takeaway is the deep relation between XAI and adversarial examples, and the hope is that this relation is clear in terms of the main ideas that can be deep dived if needed in the literature we pointed out.

7.4 Summary

- What are adversarial examples?
- Generate adversarial examples
- Transport AE from a specific ML model to a generic one
- Create universal adversarial examples
- Use AE to do XAI
- Use XAI to defend against AE

References

Adadi, A., & Berrada, M. (2018). Peeking inside the black-box: A survey on Explainable Artificial Intelligence (XAI). *IEEE Access, 6,* 52138–52160.

Fidel, G., Bitton, R., & Shabtai, A. (2020, July). When explainability meets adversarial learning: Detecting adversarial examples using SHAP signatures. In *2020 International Joint Conference on Neural Networks (IJCNN)* (pp. 1–8). IEEE.

Foolbox. (2017). *Foolbox 2.3.0.* Available at https://foolbox.readthedocs.io/en/v2.3.0/user/examples.html

Goodfellow, I. J., Shlens, J., & Szegedy, C. (2014). Explaining and harnessing adversarial examples. arXiv preprint arXiv:1412.6572.

Ilyas, A., Santurkar, S., Tsipras, D., Engstrom, L., Tran, B., & Madry, A. (2019). Adversarial examples are not bugs, they are features. In *Advances in neural information processing systems* (pp. 125–136).

Karpathy, A. (2015). *Breaking linear classifiers on ImageNet.* Available at http://karpathy.github.io/2015/03/30/breaking-convnets/

Molnar, C. (2019). Interpretable Machine Learning. A Guide for Making Black Box Models Explainable. Available at https://christophm.github.io/interpretable-ml-book/

Moosavi-Dezfooli, S. M., Fawzi, A., Fawzi, O., & Frossard, P. (2017). Universal adversarial perturbations. In *Proceedings of the IEEE conference on computer vision and pattern recognition* (pp. 1765–1773).

Nguyen, A., Yosinski, J., & Clune, J. (2015). Deep neural networks are easily fooled: High confidence predictions for unrecognizable images. In *Proceedings of the IEEE conference on computer vision and pattern recognition* (pp. 427–436).

Pearl, J., & Makenzie, D. (2019). *The book of why.* Penguin:eBook edition.

Rathi, S. (2019). Generating counterfactual and contrastive explanations using SHAP. arXiv preprint arXiv:1906.09293.

Szegedy, C., Zaremba, W., Sutskever, I., Bruna, J., Erhan, D., Goodfellow, I., & Fergus, R. (2013). Intriguing properties of neural networks. arXiv preprint arXiv:1312.6199.

Chapter 8
A Proposal for a Sustainable Model of Explainable AI

"If a lion could speak, we could not understand him."

—*Ludwig Wittgenstein*

This chapter covers:

Look at XAI full picture.
XAI in the real life: the GDPR case.
A model for XAI certification and the weaknesses of XAI.
Reflections on XAI and AGI (General Artificial intelligence).

We reached the end of this journey. In this chapter, we close the loop presenting the full picture of our point of view on XAI; in particular, we will get back to our proposed flow for XAI commenting again on it but keeping in mind all the methods we discussed.

We started the book providing impressive examples of how XAI may impact real lives; we will deep dive this aspect looking at the regulations and laws that may act as game changers enforcing or not the adoption on XAI.

We will look in some detail ad GDPR to see how to cope with it.

There is no accepted and general framework or certification to check the compliance of whatever ML model with XAI specs; we provide our point of view and envision the path to fill this gap.

Also, we deeply discuss a real case scenario in which we show how also XAI methods may be fooled to be aware of the risks of any easy approach to GDPR (or similar regulations) compliance.

We close the chapter with some speculations on XAI and AGI (Artificial General Intelligence).

L. Gianfagna, A. Di Cecco, *Explainable AI with Python*,
https://doi.org/10.1007/978-3-030-68640-6_8

8.1 The XAI "Fil Rouge"

"It's not a human move, I've never seen a man playing such a move." We started our journey like this, quoting the words of Fan Hui who was commenting the famous 37th move of AlphaGo, the software developed by Google to play GO, that defeated in March 2016 the Korean champion Lee Sedol.

We guess, or better we hope, that the readers feel in a totally different position now, trying to answer or speculating about this statement if compared with the understanding of the same in Chap. 1.

We want to look back at our XAI flow and understand what we may do, based on it, to make sense of the 37th move using this case to draw a "red line" across all the topics we learned and closing the loop (Fig. 8.1).

From a high-level perspective, AlphaGo is a Deep Neural Network that learns to play GO through reinforcement learning, becoming its own teacher. The DNN knows nothing about GO and is not trained on an existing huge dataset of GO matches but starts playing against itself combining the DNN with a powerful search algorithm.

For our XAI purposes, AlphaGo is a very complex DNN to be handled like a black box. Because of this, we cannot follow the path of intrinsic explanations that assume to have a linear model; that is not the case. Also, we cannot rely on a model-dependent approach that, as we saw, relies on the knowledge of the model internals to produce explanations.

The only viable path is the agnostic approach that handles the DNN as a black box to be interpreted. The further dilemma is to try to have a global explanation of AlphaGo behavior or get explanations on a specific case like the 37th move. This is not an easy decision.

As you can guess, AlphaGo is a very complex ML model, and just applying XAI methods like Permutation Importance or Partial Dependence Plot would not work. These methods assume that the system has been trained on a dataset that is different

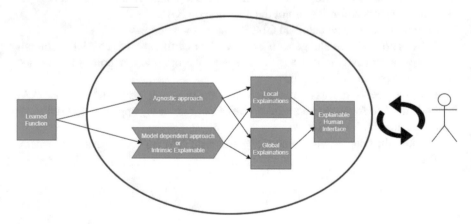

Fig. 8.1 XAI flow

from what happened for AlphaGo that learned through reinforcement learning (playing against itself). These methods work to narrow down the most important features used by the model to produce the output, but as you can guess in case of AlphaGo, even assuming to be able to produce this ranking of features, these would not be human-understandable explanations.

Our suggestion would be to go toward the agnostic approach with local explanations leveraging methods like SHAP to make sense of single outcomes like the famous 37th move. Another possible and recommended approach would be to build, as we saw in Chap. 4, local linear surrogates of AlphaGO that can be interpreted to look at some specific area of the DNN predictions (in this case, part of the match we want to explain).

As humans, the obvious question that comes out is more similar to "What would have happened if AlphaGo would not have played 37th move?"; as we saw, this means to change something that already happened in the past, but the world in which an action has not been performed does not exist because it has already passed. This is an activity of rung three of the ladder of causality that is to move toward a full explainability that is not easily achieved. In Chap. 7 we talked about example-based explanations in the context of AE and presented a possible approach following SHAP to generate counterfactuals. This would be very useful in case of AlphaGo because it would represent a way to generate human-friendly explanations to make sense of the strategy and the moves performed by AlphaGo answering contrastive questions.

Would this make AlphaGO fully explained according to our point of view? The answer is not based on what we explained in Chap. 6, playing with the damped pendulum. Following these techniques, we would not have in any case a full causal model of AlphaGO, but we may locally understand what's going on with local approximations or for single moves. Back to the analogy of damped pendulum, having the causal model means that there is no difference between the existing world in which you have the real observations of the time series for the pendulum and whatever imaginary world with hypothetical values for the same constants in the realm of knowledge discovery.

The loop is closed; starting from the question on AlphaGO, we have come back to the same one but with an arsenal of XAI methods to make sense of it coupled with the awareness of the limitations of the methods and related explanations.

8.2 XAI and GDPR

As promised in Chap. 1, we want to briefly touch the implications of regulations on data, privacy, and automated processing on the adoption of Machine Learning from XAI perspective.

We will take as an example the GDPR, the General Data Protection Regulation, that was adopted by the EU in May 2018. GDPR superseded the previous Data Protection Directive (DPD) focusing on algorithmic decision-making area.

For our scope, we are interested in Article 22 that regards automated individual decision-making (European Union 2016):

1. The data subject shall have the right not to be subject to a decision based solely on automated processing, including profiling, which produces legal effects concerning him or her or similarly significantly affects him or her.
2. Paragraph 1 shall not apply if the decision:

    (a) is necessary for entering into, or performance of, a contract between the data subject and a data controller;
    (b) is authorized by Union or Member State law to which the controller is subject and which also lays down suitable measures to safeguard the data subject's rights and freedoms and legitimate interests; or,
    (c) is based on the data subject's explicit consent.

3. In the cases referred to in points (a) and (c) of paragraph 2, the data controller shall implement suitable measures to safeguard the data subject's rights and freedoms and legitimate interests, at least the right to obtain human intervention on the part of the controller, to express his or her point of view and to contest the decision.
4. Decisions referred to in paragraph 2 shall not be based on special categories of personal data referred to in Article 9(1) unless point (a) or (g) of Article 9(2) applies and suitable measures to safeguard the data subject's rights and freedoms and legitimate interests are in place.

Paragraph 4 of Article 22 regards the treatment of personal data stating that decisions cannot be taken on the basis of personal data specified in Article 9 Paragraph 1 that are basically personal data related to ethnic origin, race, personal, and religious beliefs up to biometric data used to identify persons.

Under minimal interpretation this means that algorithms should not rely on such kind of information to take their decisions, raising a big doubt about the possibility of keeping a lot of ML models trained on such data still useful after removing the influences of these data.

Along the same path, getting closer to what regards XAI, Paragraph 3 states that a data controller "shall implement suitable measures to safeguard...at least the right to obtain human intervention on the part of the controller, to express his or her point of view and to contest the decision"; otherwise a person has "the right not to be subject to a decision based solely on automated processing."

From a legal point of view, the situation is far from being clear (Wu 2017): the right to explanations is not explicitly mentioned; GDPR only mandates that the targets of the decisions have the right to receive meaningful but properly limited information about the logic involved that is the "right to be informed." Therefore, it is fair to ask to what extent one can ask for an explanation about an algorithm. There is a lot of work in progress in this sense that could strongly alter in the next future the impact of XAI on the adoption of Machine Learning.

The ambiguity on the level of required explanations impacts the definition of any formal certification to check if an ML system is explainable or not. There is a lot of research and groups working on this from government organizations like DARPA (DARPA 2016) and European Union Commission (EPRS 2016) to big companies like IBM (IBM 2019) and Google (Google 2020).

Our point of view is that it is pretty impossible to get a kind of standard process to check explainability because it strongly depends on case to case without any possibility of an easy generalization.

8.2.1 F.A.S.T. XAI

We get back to what we stated in Chap. 1 as the minimal requirements to label a ML model as explainable: it needs to be F.A.S.T. as in *Fair* and not negatively biased, *Accountable* on its decisions, *Secure* to outside malevolent hacking, and *Transparent* in its internals.

The methods we deep dived should guide us on inspecting these specific attributes of a specific Machine Learning model. As an example, SHAP could be used to explain measures of model fairness. The idea proposed by the authors of this work (Lundberg 2020) is to use SHAP to decompose the model output among the input features and then compute the demographic parity difference (or a similar fairness metrics) for each input feature separately relying on the SHAP value for that feature.

Considering that as we saw SHAP values are meant to narrow down the main components that produce the overall model's output, we can assume that, in the same way, the sum of the demographic parity differences of the SHAP values is the top contributor to the overall demographic parity difference of the model.

This approach is tested by the authors in a case study in which the objective is to predict the risk of default for a loan. It is shown how, following this procedure, fairness bias and errors related to gender are detected by the fairness metrics.

This example covers for a practical case a possible approach to assess fairness that means to search for bias in the data, while different methods including again SHAP can be used to check the model to be accountable and transparent. Transparency doesn't necessarily mean to have access to model internals but would be enough to have the ranking of the most important features for the output and local explanations for the most interesting predictions.

A different approach is needed for security; as we saw in Chap. 7, it is pretty easy to generate AE that can be transported and universally used to attack different ML models. At the same time, we may use SHAP again, to check how the model behaves against AE or make it more robust with defensive distillation or smoothing of the function along adversarial gradients.

But we need to be aware of the fact that also SHAP or LIME can be attacked. Slack et al. (2020) show how it would be possible to fool SHAP or LIME hiding the bias of any classifier so that XAI methods would propose explanations that would not have any evidence of the bias. The authors don't just provide the theory but practically demonstrate how biased (racist) classifiers (built for the case study on the real dataset as COMPAS) would produce extremely biased results but in such a way that LIME and SHAP would not be aware of the bias. Let's go in more details about this framework that is very interesting in terms of how also XAI can be attacked.

The scenario is that an attacker, for whatever reasons, wants to deploy a biased classifier f that takes decision with impacts on the real world (loans, financial stuff). Considering that the classifier is to be certified as compliant to GDPR or similar regulations before deployment, the attacker needs to hide the bias somehow to XAI methods that will be applied to test the classifier like LIME or SHAP. Following the directions of the paper, the attacker needs to build an adversarial classifier that answers with bias on real data (the objective of the attacker) but at the same time is able to provide unbiased classifications when probed with LIME or SHAP techniques.

The core of the solution is to understand how the adversarial classifier may be able to distinguish the real inputs (to answer with bias) from the probes used to certify the model with LIME or SHAP (to answer unbiased).

As we learned, XAI methods like LIME or SHAP build local approximations of the ML model near each data point that we want to explain. The explanations can be provided because the local approximation is a linear model that is valid only in the neighborhood of the selected input. But the points generated as perturbations of the features are not necessarily belonging to the real dataset; they are synthetic points that could be out of the real distribution (OOD).

And this is exactly what we need to craft the adversarial classifier. As demonstrated by Slack et al., we can analyze the points in the original COMPAS dataset as an example and compare them to the synthetic ones generated to provide explanations with LIME or SHAP.

This analysis can be performed with a PCA to reduce the dimensionality of the full dataset containing both COMPAS and perturbed instances to 2.

The results shown in Fig. 8.2 are quite impressive.

The points coming from perturbations are distributed differently from the ones coming from the original dataset, and this can be exploited to build our adversarial classifier.

To emphasize the intuition, the adversarial classifier again would be certified as GDPR compliant (with our FAST procedure or similar one) because it would appear as absolutely fair if probed with XAI methods like LIME or SHAP. But the same classifier would be extremely biased (e.g., discriminatory) on the points coming from the COMPAS dataset.

In order to formalize this idea:

$$e(x) = \begin{cases} f(x), \text{if } x \in \varkappa_{dist} \\ \Psi(x) \text{otherwise} \end{cases} \tag{8.1}$$

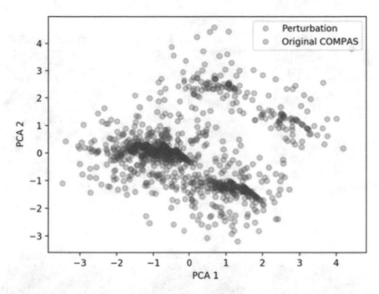

Fig. 8.2 PCA applied to original COMPAS dataset (blue points) and LIME perturbations (red points): different clusters of blue and red points naturally emerge. (Slack et al. 2020)

We consider our adversarial classifier $e(x)$ answering with $f(x)$, biased for the data point coming from the original distribution and $\Psi(x)$, a fair classifier without bias, otherwise.

In order to distinguish the points coming from x_{dist} from the perturbations and relying on our intuition (Fig. 8.1), a dataset of perturbed points (x_p) is generated to train an off-the-shelf classifier. In the specific case study of the paper, a Random Forest classifier is trained to recognize point OOD against the original. The training set is the union of x and x_p. The experimental results are very good and confirm the validity of the method to fool SHAP or LIME. The test is performed on three different datasets: COMPAS, Communities and Crime, and German Credit. For our purposes, let's look just at the results for COMPAS.

COMPAS contains information about criminal history, jail, prison time, and demographic attributes. The sensitive attribute is race (51.4% of the people are African-American). The dataset is used to have a forecast as high/low for criminal recidivism.

Figure 8.3 shows the results; it is a typical output for the ranking of features (as the ones that are used more on determining the output) according to LIME and SHAP. The biased classifier f is reported in the left column, as we see results are totally unfair because percent of African-Americans is the top feature to be used to forecast the recidivism. But according to LIME or SHAP, the unbiased function Ψ (as shown in columns 2 and 3) uses 1 or 2 features (that are uncorrelated with sensitive attributes) for predictions. It is important to stress that Ψ is not to be restricted to use always uncorrelated features. It can also use existing features in the dataset to make predictions.

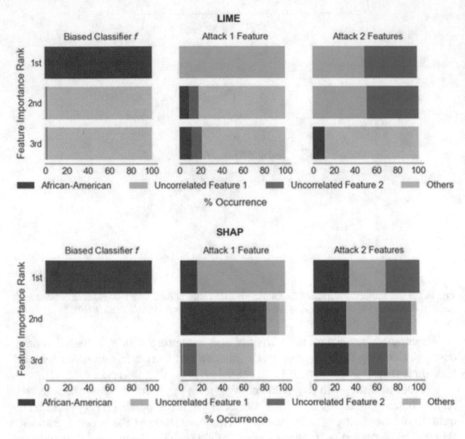

Fig. 8.3 Feature importance for the biased classifier vs our adversarial classifier (middle and right columns) that exhibits an unbiased behavior for the data points out of the original distribution (perturbation). (Slack et al. 2020)

In the paper there is also an evaluation of the fidelity of e if compared to f in terms of the percentage of the instances on which their predictions agree. The most important point to have an effective adversarial classifier is an accurate OOD detector to discriminate points coming from the original distribution vs the perturbed ones that LIME and SHAP use to get the most important features.

This kind of risks of being unable to use XAI to determine if a ML model is behaving fairly or not is a big deal and as remarked by Merrer and Tredan (2019) who claimed to have the proof of the impossibility for online services to provide trusted explanations.

And as you can easily guess, online services are everywhere in our daily life from Facebook to Google. Merrer and Tredan (2019) submitted the paper to "Fairness, Accountability, and Transparency" Conference to show the problems related to XAI and the difficulties in trusting online services.

They used the metaphor of a bouncer in a night club. You are denied entering, and the bouncer can answer whatever reason for this decision: not the right dress or shoes. But you cannot be sure that it is not making the decision based on your race. And an algorithm can behave exactly in the same way as we have shown in the previous example; an algorithm could also be looking fair against XAI probing.

A way to detect the bias of the algorithm would be to rely on incoherent pairs that are found in the metaphor of the bouncer – someone that has been blocked because of a red skirt and then someone that entered dressing a red skirt. But it is not like that with an algorithm working on thousands of features.

An alternative would be to check the ML models on premise and not to rely just on agnostic methods probing the online instance from outside as a black box. Otherwise, the algorithm, as we saw, may change the answers depending on the source of requests.

As argued by the authors, this is not so different from hygiene inspections in which it is impossible to check the restaurants by just checking the dishes that are served – you need food inspectors into the kitchen to control the practices and equipment that are used to serve those dishes.

This closes our journey: at the moment in which we are writing, there is not yet real enforcement of effective regulations to guarantee XAI, but whatever the regulations should be we tried to share the awareness that also the adoption of XAI methods may not be enough to certify a ML system as Fair, Accountable, Secure, and Transparent. We dedicate a short appendix (Appendix A) to envision an executable flow to go through a certification of a ML system as "Explainable" (we name it F.A.S.T. XAI certification), it is based on what we learned in this book and provides an operational approach to assess and analyze the ML "opaque" systems to get explanations.

And we close just in time with the last section with some thoughts on GAI, XAI, and quantum mechanics (yes theoretical physicists like us need to talk at least in few lines of quantum mechanics whatever the topic of the book is).

This closes our journey: at the moment in which we are writing there is not yet real enforcement of effective regulations to guarantee XAI, but whatever the regulations should be we tried to share the awareness that also the adoption of XAI methods may not be enough to certify a ML system as Fair, Accountable, Secure and Transparent. We dedicate a short appendix (Appendix A) to envision an executable flow to go through a certification of a ML system as "Explainable" (we name it F.A.S.T. XAI certification), it is based on what we learned in this book and provides an operational approach to assess and analyze the ML "opaque" systems to get explanations.

8.3   Conclusions

"If a lion could speak, we could not understand him and not because of different languages but because of two different worlds or better two different 'language games'." We open this last section quoting Wittgenstein with his famous statement on language games and the limits of the language.

It is interesting to look at this statement from an AI and XAI perspective. All the ML models we tried to explain with XAI belong to "Weak AI," that means a kind of intelligence that is different from the human one and is specialized in solving specific tasks.

In contrast to the "Weak AI," we should consider the "Strong AI" that is the attempt to realize an intelligent agent indistinguishable from the behavior that a human would show in the same situation.

The term "Strong AI" was born in the context of Searle's work in 1960 and the famous mental experiment called "Chinese Room." Suppose we have created an agent that would behave as if he understood Chinese; this means that the agent runs a program that allows the manipulation of the received inputs (Chinese characters) so to provide answers in Chinese and to pass a hypothetical Turing test being indistinguishable from a human person who understands Chinese and answers the same questions. Would we say that this Artificial Intelligence really understands Chinese?

While we are writing, OpenAI team released GPT −3, the last language model, in May 2020 (OpenAI 2020), and this model has been used to generate poetry, write adventures, or create simple apps with a few buttons. GPT −3 is avowed as one of the best candidates to be classified as AGI (Artificial General Intelligence) that is another way to say "Strong AI" and means to have an artificial agent that is capable to perform any task as well as a human being.

The interesting point about GPT −3 is that there is no disruptive innovation in terms of architecture or theoretical approach if compared to the state of the art of NLP.

The main difference is scale: GPT −3 had 110 million parameters, while GPT −2, in its largest iteration, had 1.6 billion parameters. The consensus is that this scale-up is making the real difference for GPT −3; the impressive improvement is that GPT −3 is able to handle the task of few-shot learning, the "Holy Grail" of human intelligence. A kid doesn't need to see thousands of cats to recognize a cat, few cats are enough, and this is pretty different from the standard learning needed by DNN. GPT −3 is able to complete sentences without specific learning, and more than that it can do many other tasks like translate between languages, perform reading comprehension tasks, or answer SAT style exam questions without any specific training on each specific task.

This has to be compared with the case of AlphaGO that is not able to even play tic-tac-toe or checkers albeit pretty easy games if compared to GO.

But the question remains the same: does GPT −3 or any similar AGI system really understand the meaning of what it is doing? Is a "Chinese Room" kind of experiment enough to check this out? This is the domain of philosophy, and thousands of books may be cited just to clarify the word "understanding." The argument of scale is intriguing considering that a human brain has roughly 100 billion neurons, which forms something of the order of 100–500 trillion synaptic connections. If scale is the solution to humanlike intelligence, then GPT −3 is still about 1000x too small. But to be fair, we should also add that a human brain consumes about 20 W to be compared to the enormous amount of energy required by GPT −3 or similar agents.

But for our purposes and from perspective of XAI, we ask a different question: "Should we enforce explainability on agents like GPT -3?" or better "Should we add the strong requirement of being able to produce human-understandable explanations to define an agent as AGI?"

The answer is not simple. From one side, the strong requirement of being explainable naturally emerges on the basis of what we discussed (think about human in the loop); as we saw, the true understanding of a system passes through rung 3 of the causal ladder, that is, the rung of human kind of activities like imaging, counterfactuals, and knowledge discovery.

So, we could not label an agent as AGI without the capability of climbing by itself up to rung 3. But at the same time, the words of Wittgenstein and related stream of research sound like an alarm: Are we sure that we could really understand an AGI agent that lives in a different world? Would it be a kind of lion for us that we would not understand also assuming it would speak? Are we constrained to live in different language games as humans and AGI agents?

But as authors we cannot forget our theoretical physics background. Quantum mechanics can be considered one of the best and beautiful theories we have on the physical world. It is confirmed by every experiment, and the applications on everyday life are impressive: from our PCs and mobile phones to laser and communications, from transistors to microscopy and medical diagnostics devices. But despite all these, the stream of research that still struggles with *interpretation* of quantum mechanics is very active and prolific. And we don't use the word "interpretation" by chance.

We can call quantum mechanics a physical theory, but as any other physical theory or model, it can be considered as a computational tool: given some input, the theory prescribes how to calculate the output, how to get the predictions on the system we are studying in terms of how it behaves.

Quantum mechanics provides impressive predictions on its domain, but we are not yet able to cope with its interpretation: the existence of probability and uncertainty at the very root of the theory for any physical observable undermines our attempts as humans to make sense of it; to quote Mr. Feynmann: "If you think you understand quantum mechanics, you don't understand quantum mechanics."

We attempted different approaches to deal with this state of things, from hidden variables to many world interpretations; the objective is to avoid to take the uncertainty, the superposition of states, and wave-function collapse to be real in order to have a more classical and deterministic interpretation of physical observables.

But up to now, there is no definite success in this sense, and we continue on using QM albeit a lack of a satisfactory and final interpretation. The alternative is to take it as it is, as a predictive tool, and look at quantum mechanics as a theory of relations (relational quantum mechanics, RQM).

The physical variables don't describe the "things": they describe how the things interact with each other and the state of a quantum system as being observer-dependent. So as from the beginning, with Galileo's inertial frames and then special and general relativity, there is no sense to talk about the "true" phenomenon or physical event, but what is experienced depends on the observer: the properties of an object that are real with respect to a second object are not necessarily the same with respect to a third object. We exit from a world of "things" to go into the "world" of interactions where what we see are just interactions, not tangible things like stones.

What is the relation of this with XAI? The metaphor is that we may look at ML in such a way: as a tool to get impressive predictions that we have to double check. ML can be considered a computational tool like QM that, given an input, produces an output. Because of this analogy with RQM, we should not be too rigid in terms of the expected interpretations and explanations. We can get explanations with XAI, but we might need to accept that the same explanations may depend on the type of interactions we have with our ML models. The ML models may work fine and rely on inner computation and paths that we may be forced to accept as intrinsically different from how we would work as human beings.

Also, the explanations are results of interactions, and taking the road to AGI seriously, we, as human observers, might need to relax our requirements on rigid and strong explanations. Like in a conversation with a lion, we should accept that the explanations are not just out there, ready and available in the form we expect; these explanations may belong to a different language game and may be results of new type of interactions in which everything depends on the relations between the entities and there is nothing that is existent by itself in a fixed form.

There is nothing substantial or persistent out there in the universe, just an infinite net of interactions and dynamical processes in which we are not privileged observers.

8.4 Summary

- Adopt XAI methods to interpret complex ML models like AlphaGO.
- Adopt FAST criteria to check GDPR compliance.
- Use SHAP to check fairness.
- Learn how XAI methods can be fooled by adversarial attacks.
- Understand what is AGI and the possible implications for XAI in the next future.

References

DARPA. (2016). *Explainable Artificial Intelligence (XAI)*. Available at https://www.darpa.mil/program/explainable-artificial-intelligence

European Union. (2016). *GDPR*. Available at https://europa.eu/european-union/index_en

EPRS. (2016). *EU guidelines on ethics in artificial intelligence: Context and implementation*. Available at https://www.europarl.europa.eu/RegData/etudes/BRIE/2019/640163/EPRS_BRI(2019)640163_EN.pdf

Lundberg, S. (2020). *Explaining measures of fairness with SHAP*. Available at https://github.com/slundberg/shap/blob/master/notebooks/general/Explaining%20Quantitative%20Measures%20of%20Fairness.ipynb

Google. (2020). *Explainable AI*. Available at https://cloud.google.com/explainable-ai

IBM. (2019). *Introducing AI Explainability 360*. Available at https://www.ibm.com/blogs/research/2019/08/ai-explainability-360

Merrer, E. L., & Tredan, G. (2019). The bouncer problem: challenges to remote explainability. arXiv preprint arXiv:1910.01432.

OpenAI. (2020). *OpenAI API*. Available at https://openai.com/blog/openai-api/

Slack, D., Hilgard, S., Jia, E., Singh, S., & Lakkaraju, H. (2020, February). Fooling Lime and Shap: Adversarial attacks on post hoc explanation methods. In *Proceedings of the AAAI/ACM Conference on AI, Ethics, and Society* (pp. 180–186).

Appendix A

"F.A.S.T. XAI Certification"

The purpose of this checklist is to provide a practical guidance, based on the contents of this book, and the steps to perform to conduct an assessment and a possible XAI certification of a ML system.

The main idea is that *before* using pure XAI methods in (6), we need to build the questions we want to be answered otherwise we would invalidate the XAI methods themselves.

Another critical step is step (4) in which we ask for a surrogate model, this would guarantee the better performance of the model in respect to the baseline model (the surrogate) and also would add confidence in terms of lack of bias. Step (4) is somehow optional, without it we would get a lighter form of certification.

Accountability in step (5) is strongly related to the legal aspects and may change depending on the specific regulation that is in place (e.g., GDPR).

1. *Model preparation*

    – [ ] Acquire the model to explain (a.k.a the *"MODEL"*)

2. *Frame the problem*

    – [ ] Data source identification
    – [ ] Identify sensible and possible adversarial features

3. *Construct the "What If?" counterfactual scenario*

    – [ ] Compile a list of possible counterfactual questions adapted to realistic usage scenarios

© The Author(s), under exclusive license to Springer Nature Switzerland AG 2021
L. Gianfagna, A. Di Cecco, *Explainable AI with Python*,
https://doi.org/10.1007/978-3-030-68640-6

4. *Build a surrogate model (a.k.a. the "SURROGATE")*

  - [ ] If data are openly available train an intrinsic global explainable model, the "SURROGATE" else ask for a surrogate model to the authors
  - [ ] Explore feature importance on the surrogate model

- If the "SURROGATE" is not available, certification is assumed to be "LIGHT"

5. *F.A.S.T. methodology key aspects*

  - [ ] Identify all possible fairness problems in data [F]
  - [ ] Qualify the accountability aspects of the model [A]
  - [ ] Describe the security of the data [S]
  - [ ] Leverage transparency of "SURROGATE" model [T] (not needed for LIGHT certification)

6. *Directly explain the real model*

  - [ ] Train a XAI model directly on the "MODEL" we want to analyze
  - [ ] Display the importance of global and local features
  - [ ] Assure that "What If?" question can now correctly be answered

Index

Printed in the United States
by Baker & Taylor Publisher Services